# Biotechnological Tools for Increasing Productivity of Pulse Crops

## ABOUT THE BOOK

The application of recombinant DNA technology has revolutionized plant breeding in recent years. Genetic transformation in plant breeding greatly increases the gene pool and broadens the scope of genetic changes that modern breeders may draw upon. The first generation of genetically modified (GM) crop plants hold great promise in carrying new or improved agronomical traits that require less intensive farming methods or which have improved quality traits. These GM crop plants have been produced to improve farming production systems, aiming, for instance, to reduce negative impacts on the environment, such as fertilizers, herbicides and pesticides. Genetically modified plants are created by the process of genetic engineering, which allows scientists to move genetic material between organisms with the aim of changing their characteristics. All organisms are composed of cells that contain the DNA molecule. Molecules of DNA form units of genetic information, known as genes. Each organism has a genetic blueprint made up of DNA that determines the regulatory functions of its cells and thus the characteristics that make it unique. Prior to genetic engineering, the exchange of DNA material was possible only between individual organisms of the same species. With the advent of genetic engineering in 1972, scientists have been able to identify specific genes associated with desirable traits in one organism and transfer those genes across species boundaries into another organism. This book on gene biotechnology is being prepared by keeping in view the requirement of the students and researchers in biotechnology, genetic engineering and other allied fields.

## ABOUT THE AUTHOR

**Dr. Uday Kumar**, Presently Asst. Professor at Janki Devi Mahavidyalaya, Darihat under B. R. Ambedkar University, obtained all his degrees M.Sc and Ph.D. in biotechnology. He is teaching this subject for last thirteen years, since 2005. He has published three research papers and attended Nine National level seminars and presented research papers. His dynamism is also seen in various other fields apart from his successful academic career.

# Biotechnological Tools for Increasing Productivity of Pulse Crops

**Dr. Uday Kumar**

**AMIGA PRESS. INC**
**Delhi -110051**

**AMIGA PRESS INC.**

H.O :- 28-A, Indra Park,

Chander Nagar, Delhi :- 110051 (India)

Tel :- 011-22504028, +91-98736-57000

Email :- amigapress@gmail.com
amigaeditorial@gmail.com
amigapub@gmail.com

Website :- www.amigapress.co.in

This edition is published by Amiga Press Inc.

ISBN : 978-93-87295-39-1



First Published 2019



---

Printed and bound by Replika Press Pvt. Ltd.

# Preface

A transgenic organism is an organism which has been modified with genetic material from another species. The genetic modification is accomplished by inserting DNA into an embryo with the assistance of a virus, a plasmid, or a gene gun. The embryo is allowed to develop, and the mature organism will express the DNA which has been inserted into its genome. Transgenic organisms can also pass the modification on to future generations by breeding with other members of the same species. Organisms may be genetically modified for any number of reasons.

Biofortification of food crops makes sense as part of an integrated food-systems approach to reducing malnutrition. It addresses the root causes of micronutrient deficiencies, targets the poorest people, and is scientifically feasible and cost-effective. It is a first step in enabling rural households to improve family nutrition and health in a sustainable way. HarvestPlus is a coalition of CGIAR Future Harvest Centres or Institutes, partner collaborating institutions and supportive donors. The International Centre for Tropical Agriculture and IFPRI are coordinating the plant breeding, human nutrition, crop dissemination, policy analysis and impact activities to be carried out at international Future Harvest Centres, national agricultural research and extension institutions, and departments of plant science and human nutrition at universities in both developing and developed countries.

Farmers and pastoralists have manipulated the genetic make-up of plants and animals since agriculture began more than 10 000 years ago. Farmers managed the process of domestication over millennia, through many cycles of selection of the best adapted individuals. This exploitation of the natural variation in biological organisms has given us the crops, plantation trees, farm animals and farmed fish of today, which often differ radically from their early ancestors.

The aim of modern breeders is the same as that of early farmers-to produce superior crops or animals. Conventional breeding, relying on the application of classic genetic principles based on the phenotype or physical characteristics of the organism concerned, has been very successful in introducing desirable traits into crop cultivars or livestock breeds from domesticated or wild relatives or mutants. In a conventional cross, whereby each parent donates half the genetic make-up of the progeny, undesirable traits may be passed on along with the desirable ones, and these undesirable traits may then have to be eliminated through successive generations of breeding. With each generation, the progeny must be tested for its growth characteristics as well as its nutritional and processing traits. Many generations may be required before the desired combination of traits is found, and time lags may be very long, especially for perennial crops such as trees and some species of livestock. Such phenotype-based selection is thus a slow, demanding process and is expensive in terms of both time and money. Biotechnology can make the application of conventional breeding methods more efficient.

Broadly speaking, biotechnology is any technique that uses living organisms or substances from these organisms to make or modify a product for a practical purpose. Biotechnology can be applied to all classes of organism-from viruses and bacteria to plants and animals-and it is becoming a major feature of modern medicine, agriculture and industry. Modern agricultural biotechnology includes a range of tools that scientists employ to understand and manipulate the genetic make-up of organisms for use in the production or processing of agricultural products.

Some applications of biotechnology, such as fermentation and brewing, have been used for millennia. Other applications are newer but also well established. For example, micro-organisms have been used for decades as living factories for the production of life-saving antibiotics including penicillin, from the fungus Penicillium, and streptomycin from the bacterium Streptomyces. Modern detergents rely on enzymes produced via biotechnology, hard cheese production largely relies on rennet produced by biotech yeast and human insulin for diabetics is now produced using biotechnology.

This is a reference book. All the matter is just compiled and edited in nature, taken from the various sources which are in public domain.

This book on gene biotechnology is being prepared by keeping in view the requirement of the students and researchers in biotechnology, genetic engineering and other allied fields.

*—Editor*

# Contents

# 1

# Agriculture and Transgenic Crops

Genetic engineering is an ever-expanding field in the world of science. In the last ten years, the process of genetically altering organisms has evolved from experimentation to commercial application. It is taking the place of selective breeding for isolating desired traits in different organisms. Transgenics has also led to a better understanding of gene function, allowing for better understanding of human diseases such as Alzheimer's, diabetes, and cancer. For some, such as the starving poor and those afflicted with genetically-caused diseases, transgenic organisms are a blessing. For others, such as poor farmers and the animals themselves, genetic engineering spells the end of the world.

The Federation of European Laboratory Animal Associations defines the term transgenic animal as an animal whose genome has been deliberately modified (Margawati). This definition also holds true for any other transgenic organisms, including plants and bacteria.

In fact, the principle of transgenics is the introduction of an alien gene into an organism's DNA. Generally, the gene would not occur naturally in its host organism.

Transgenic science was born in 1970 when experiments were conducted to produce different strains of mice; cells originating from one strain were introduced to the embryos of another strain by aggregation or injection during the blastocyst stage (C. River). The resulting offspring showed characteristics from each strain used. These first transgenic animals were created without the use of recombinant DNA, but the process was a building block for the mechanisms of genetic engineering used today (C. River). This process is actually quite similar to that of cloning, when one organism is a genetically identical to its parent.

## FRUSTRATE TRANSGENIC AGRICULTURE SYSTEM

Although there are legitimate reasons to oppose genetic engineered agriculture, or at least to demand the most careful controls, there are a community of opponents who have taken their opposition beyond what is ethical. I am not talking about a reasonable opposition expressing the concerns summarized earlier, but rather about an opposition for which the ends justify the means, including lies, vandalism, etc. I need to stress this point. There are many people, sincerely opposed to genetic engineered crops, whose ethics I do not question. I believe that most of the opponents of biotechnology would fall into that category. Yet, in too many cases, those sincere concerns are based entirely on misinformation which originated in deliberate lies and fear mongering.

We are here exposing the ethics of the people who have created the lies. These opponents must have some motive, and it seems that an alliance has emerged between at least four groups, each with its own agenda. First there are people who sincerely believe that genetic engineering is ethically wrong, and that anything they do to stop it from happening is therefore right. Second there are foreign governments and their constituencies who are worried about American domination of agriculture. Closely related to this are advocates of organic agriculture who seem to be engendering public fear to make their own products more salable.

Third, there are environmental groups who have been misled by a radical fringe and have become willing to do anything to stop genetic engineering agriculture. The most conspicuous among these groups is Greenpeace. Finally, there are people opposed to capitalism or to large businesses dominating agriculture. Opposition from environmental groups is particularly frustrating to me. Most measures which benefit the environment require people to give up something, and they don't like to do that. Recycling is nearly painless and saves money, but many people won't make the effort. A few degrees adjustment of a thermostat could save vast amounts of energy, but most people would rather be comfortable. We end up settling for half measures. But here is a technology that benefits the environment without asking people to give up anything, and its biggest opposition comes from environmental groups.

### Misinformation about Food Safety

There is a deliberate campaign to frighten people about the safety of the food supply. This campaign has worked successfully in England and in much of Europe. Dr. Arpad Pusztai, who worked at the Rowett Institute in Aberdeen,

Scotland, performed an experiment. It began when a gene was transferred from a poisonous plant, the snowdrop, into a potato. The transferred gene specifies the production of a poisonous compound called lectin. Dr. Pusztai proceeded to experiment with rats. Some rats were fed with the raw potatoes which were genetically engineered to contain the poison. The rats in the control group were fed ordinary raw potatoes and were also given the amount of lectin poison which the first group of rats would have gotten from eating the transgenic potatoes. Both groups of rats developed malformed organs, and there was no statistically significant difference between the rats who consumed the poisonous potatoes and those who consumed the poison.

However, Dr. Pusztai claimed that his data showed that the rats who ate the genetically modified potatoes had more deformed organs. No scientific journal would publish Dr. Pusztai's interpretation, and his institution would not support him. He hired an independent statistician to review his data, who also considered the data to show no difference between the two groups. But Dr. Puzstai held to his opinion. Eventually the disagreement became serious enough that his connection with the Rowett Institute was ended. The opponents of genetic engineering, mostly in England, have blown this result into a cause celebre. Dr. Pusztai is portrayed as muzzled by the scientific establishment, although the British medical journal Lancet eventually published Pusztai's paper over the recommendations of its reviewers because of the widespread public interest.

The British tabloid press covers this story continuously, with lurid photographs of deformed rat organs. The potatoes genetically engineered to be poisonous became synonymous with all transgenic food, called Frankenstein food in the tabloids. There are varieties of potatoes, bred to be eaten, engineered to resist insects, viruses and fungi. All these varieties have been fed to rats and have never harmed them. The opponents of transgenic food have no explanation for that - they are content to use one probably misinterpreted experiment with potatoes nobody will ever eat, to stir up doubts about food safety. During safety testing of transgenic tomatoes, Dr. Belinda Martineau discovered that when rats are fed huge amounts of tomato paste they can develop stomach lesions.

It doesn't matter whether the tomatoes are transgenic or not. But before there were transgenic tomatoes, nobody had ever fed rats such large doses. Martineau's results were reported to the FDA but opponents routinely call this a "cover-up" of a health hazard with transgenic tomatoes. As another example, I read an op-ed article in the Boston Globe in August of 1999, written by Paul Billings, a member of the board of the Council for Responsible Genetics. The gist of the article is that dangerous untested foods are being foisted upon an unsuspecting American public, by mad scientists. Every genetically engineered crop has been tested for safety. The testing has been much more extensive than that for any other foods, including foods developed by radiation induced mutation. Dr. Billings knows this. He knows about the government testing rules that establish the safety of each individual crop.

He wants genetically modified food to be tested as strictly as drugs are tested. Billings' op-ed article contains only one 'fact' that most people would not have known before - the rest is either his opinion or just wrong. He says that transgenic soybeans have been shown to be deficient in a certain unidentified nutrient. It is not easy to track down the source of this 'fact', but I did it. The nutrient in question is a phyto-estrogen (also known as a phytosterol). Although phyto-estrogens are not essential to human health, there is some indication that they help prevent cancer. The study that indicates that transgenic soybeans are deficient in phyto-estrogens comes from Dr. Marc Lappe, who wrote the book "Against The Grain", a polemic against genetic engineering and especially against the Monsanto Co., the leading company in the field, which developed the soybeans in question. Here is how Dr. Lappe established that genetic engineered soybeans are defective. Understand that there are dozens of varieties of soybeans with the herbicide tolerance trait and over a hundred varieties of conventional soybeans. Dr. Lappe compared one conventional variety with one transgenic variety. He found a 12 per cent difference. But individual soybean varieties vary by more than 100 per cent in their phyto-estrogen content.

Also phyto-estrogen content is not stable. It declines with storage, by much more than the 12 per cent difference. But Lappe at least reports his data along with his biased interpretation of it. Dr. Billings reports only the interpretation without any indication that it comes from a biased scientist whose own data show insignificant variations in a nutrient whose role in human health is not even firmly established. Dr. Billings seeks only to mislead people. Not one reader in a thousand would do what I did, track down the data.

His purpose is to plant a little seed of doubt about food safety, hoping that it will fester in our minds, mingle with similar misinformation, and eventually become accepted fact. Not everything misleading is false. Propagandists are very skilled at making true statements that seem to imply something quite different.

For example, they frequently say that the FDA's rules about testing genetically modified food are voluntary, implying that some testing doesn't get done. Under current law, FDA has no authority to require safety tests for any food, transgenic or conventional, although it can prevent the sale of foods it considers unsafe. But, in fact, each developer of a transgenic crop has consulted with FDA and performed every test FDA suggested. They also like to say that safety test results are trade secrets. True, it would be legal to keep test results secret.

But no developer has done it. Earlier I mentioned a problem with a soybean with a Brazil nut gene. People allergic to Brazil nuts should not expect to have to avoid soybeans, but the allergen was identified by testing and therefore the modified soybeans were never created.

Also, the scientists who demonstrated this published their results in a scientific journal, and this led the FDA to not allow any gene to be transferred into a food from a species known to cause allergies. That should be seen as evidence that the genetic engineers are responsible people, and that testing is working well. But the unethical opponents of biotechnology routinely present this episode, carefully worded, as if there was a near disaster, revealing gross problems with the current regulatory system.

## MORE FROM THE SUN

Plants derive energy from sunlight and use it to make sugar from carbon dioxide and water. This is photosynthesis. Scientists still do not have a complete understanding of how photosynthesis happens, although they know most of the steps. They know many of the genes which create the proteins needed for photosynthesis. They also know that there are differences in photosynthesis from one species to another. It happens that corn is an overachiever.

Corn plants make more sugar per unit of sunlight than any of the other grains. An international team of scientists from Japan and from Washington State has transferred three of corn's photosynthesis genes into a rice plant. Early indications are that the transformed rice is more productive than the original rice variety. A more important potential application may be the development of very fast growing trees. If global warming cannot be prevented by adding less carbon dioxide to the atmosphere, by burning less coal and oil, the only alternative is to depend on processes that remove it. Number one on that list is growing new trees.

Anything that makes agriculture more efficient can make more land available for growing trees. Anything that makes those trees grow faster removes more carbon dioxide from the atmosphere. For many environmentalists, preventing global warming is the highest priority. But proposed measures to restrict burning fossil fuels have encountered fierce political resistance. Opponents claim that such restrictions would be cost too much money, and would cost people their jobs, their comfort and their prosperity. By contrast, nobody loses anything if carbon dioxide is removed from the atmosphere by growing new trees.

## *Toxic Soils*

Some soils are poor for plant growth because their mineral content is toxic. A high aluminum content is the most frequent problem, especially in acidic soils. But it has been possible to identify a few genes which enable some plants to extract aluminum compounds from soil and sequester them harmlessly in their fibrous parts. Recently, Florida scientists discovered a type of fern which can extract arsenic from the soil, although they do not yet know how the fern does this. But other teams have identified genes that can enable plants to remove cadmium, zinc and mercury from soils. By transferring such genes to fast growing plants, it should be possible to clean up some toxic soils in much the same way as we can use bacteria to clean up oil spills. In the nearer term, there is the work of Mexican scientist Luis Herrera Estrella. He transferred into corn a gene that allows the plant to overproduce a natural chemical, citric acid, which it then excretes through the roots. Citric acid binds to aluminum and prevents the plant from taking it up from the soil. Herrera's approach is not to extract aluminum from the soil but to prevent it from passing from the soil to the plant. A much larger problem is salt-contaminated soil caused by irrigation. Rainwater is very pure, but water borrowed from rivers contains some dissolved salt. Over many years of irrigation, the salt accumulates. But water cannot get from soil to roots if the soil water is saltier than the intracellular water.

In fact, water goes the other way, from plant to soil, and the plant dies. A gene was identified in a relative of cabbage. This gene enables the plant to pump salt from the soil into an isolated part of a cell, called a vacuole, where it is stored without harm to the plant. When salt is thus removed from the soil around the roots the plant can then take up the less salty water. The salt-tolerance gene was experimentally transferred to a tomato plant, where a control gene keeps it turned on all the time. The resulting tomato plant is able to grow well in salty soils. Happily, the fruit is not high in salt, but the plant's stems, leaves and roots are loaded with salt, so after the growing season the plant parts could be shipped elsewhere, making the soil become less salty each year. It's one more case of an environmental problem that can be solved by gene transfer.

## *Biological Pest Controls*

Farmers, for very obvious reasons, would prefer not to use pesticides. They cost money and they are dangerous to use. Farmers much prefer Integrated Pest Management (IPM), a system that combines many different methods of suppressing crop pests, including encouraging predatory insects. Farmers even buy them. Agricultural distributors can supply such insects as ladybugs, praying mantids, lacewings and parasitic wasps. Integrated pest management includes using pesticides when other means are insufficient. But when a crop is sprayed with conventional insecticide, the harmful insects are not the only victims. Predatory insects may also be wiped out. Without any predators available, the pest populations can recover quickly, so that a second application of pesticide is required, which also kills the insect predators. This vicious cycle could be broken if genetical engineers can develop predatory insects resistant to common pesticides.

## *Rust Resistance*

To a plant scientist, rust has nothing to do with oxidized metal. It is a plant disease caused by a fungus. It blights all the cereal crops, barley, wheat, oats, corn, millet and sorghum, but not rice. The rust fungus reproduces itself by forming club shaped cells called basidia. Each basidium bears four spores. When the spores are ready, they are released and carried by the wind. The fungus infecting a single grain of wheat can easily produce millions of spores. The spores are so light that they can travel several times around the world before falling to the ground. Although some varieties are more resistant to rust than others, no variety is immune. But rice must contain some combination of genes that confers immunity to rust. If these genes can be identified and if their function can be deciphered, it should be possible to transfer them to other cereal crops and end, once and for all, this most important cause of famine.

## *Fast Growing Trees*

Making paper requires large amounts of natural cellulose. Some can be derived from recycling, but most of our paper is made from freshly cut trees. The best trees for paper-making are fast growing softwoods with low resin content, like aspens. Genetic engineers have transferred genes for pest resistance and herbicide resistance into aspen and have tinkered with the genetic switches that promote growth to create a fast growing aspen that could supply our paper needs using considerably less land. The paper-making process must bleach out the brown colour of lignin, one of the components of wood. The bleaches used to be dumped in the nearest rivers, an important and highly visible kind of pollution. This is no longer allowed, but the disposal of chemicals from paper mills is still a major headache. At the Michigan Tech University, researchers have reduced the lignin content of aspen so that fewer chemicals are needed in the paper making process.

### *Fast Growing Fish*

Most of the salmon we eat are caught wild, but some are grown in farm ponds. It takes about three years for a salmon to grow from fingerling size to optimum marketing size. In wild salmon a control gene turns on the gene for growth hormone, but only in the pituitary gland and primarily in warm water. So genetic engineers used a different control gene to turn on a growth hormone gene in cold water. That control gene was transferred from an ocean pout, and it originally turned on a gene for a protein that helped the pout tolerate very cold water. The resulting creature looks and tastes just like the wild type salmon but it grows three times faster so it ought to be cheaper to produce. Wild salmon are now under environmental pressure from overfishing and because many of the streams where they lay their eggs are either polluted or inaccessible. If farmed salmon can economically replace more wild salmon, the pressure on this desirable species could be reduced dramatically. There's a need for fast growing fish in rice growing regions. Rice is planted in standing water, but it is harvested from dry ground. With a slow growing rice farmers often raised fish in the rice paddy alongside the young plants. But newer strains of rice mature almost twice as fast as traditional varieties. Although this lets farmers grow more crops per year, unfortunately the rice paddies are not flooded long enough to raise fish. If genetic engineers were able to make fish grow faster, the farmers could again exploit this valuable protein resource.

### *Consumer Traits*

Most of the traits in the examples mentioned so far have been targeted at the producer. The no-till soybeans are cheaper to grow because ploughing costs money. The cotton is cheaper to grow because chemicals cost money. The chymosin from yeast is cheaper than chymosin from calves. The consumer doesn't know how much water or fertilizer the farmer used, or whether a salmon is one year old or three. The vitamin enriched rice is the only one of the examples where the final product is better, rather than cheaper, for the person who eats it. But in the future we can expect to see "consumer traits". One of the first to appear will be potatoes genetically engineered to have a higher percentage of solids.

If you love french fried potatoes but don't like the calories, you will love the new potatoes. They will absorb less oil but stay crispier longer. Another valuable trait coming soon is coffee beans without caffeine. Caffeine is now removed from coffee by a chemical treatment, invented by German chemist Ludwig Roselius. Decaf is the only coffee I drink so I am looking forward to cheaper decaf coffee. In addition, many common foods are not safe for everyone.

For example, peanuts cause a life threatening allergy in some people, especially children. Allergens are unusual proteins which are digested very slowly. Someone who has an allergy to a widely used food needs to read the small print on product labels, quiz the waiter in a restaurant, etc. If a child has the allergy, the problem is that much harder to manage. But peanuts or other crops are now being genetically engineered to eliminate these allergens. USDA scientists have identified the principal allergen in soybeans and have successfully developed modified soybeans which do not produce that allergen. It is much easier to deactivate a gene, once its function is discovered, than it is to transfer a gene from one organism to another. The same antisense trick that was used to delay ripening can be used to suppress synthesis of an allergen. Therefore, once a gene has been identified which codes for an allergenic protein, the technology to eliminate that allergen from food crops is relatively easy, unless the allergenic protein is important to the life processes of the plant.

### *Lawns Don't Need Frequent Mowing*

In addition to food and fibre, genetic engineers are working to modify grass. Farmland constitutes mankind's biggest footprint on the earth, but lawns, athletic fields, golf courses, etc. have the biggest footprint in some communities. Environmentalists have many criticisms of lawns. They must be mowed regularly, which uses gasoline, creates noise pollution, and takes up people's time. In addition, the growing grass uses a very significant amount of water, fertilizers and pesticides, to make longer leaves which are then cut off by the mower. But genetic engineers are trying to develop a variety of grass which reaches a desired length and then dramatically slows its growth. Combined with pest resistance genes, this new grass would be almost as simple to maintain as astroturf.

## CROSSING OVER IN PLANTS

Plants and animals have two of each gene, on separate but similar chromosomes, analogous to having two copies of the document. The two copies are usually slightly different. By analogy, if the document were The

Lord's Prayer, one copy might read The Lord is my shepherd! I shall not want. The other copy might read God is my guide! He provides for me. In sexual reproduction, the two documents are sliced apart and put back together, so that they might read The Lord is my guide! He provides for me. and God is my shepherd! I shall not want. This process is called crossing over. This happens in the sex cells of both parents and then the offspring inherit one "document" from each parent. Crossing over is a major source of genetic variability. Sometimes, however, it produces the same kind of nonsense as in the example "don't be happy! drink the water!".

### *Transposons*

Genes are arranged on the chromosome in order but many genes make a habit of jumping to another part of the chromosome. They are routinely snipped out of the DNA and put back in a different place. These jumping genes are called transposons. By being moved to a different part of the DNA, the functional expression of a transposon changes. A familiar example of a transposon is a gene that determines the colour of the kernels of Indian corn.

### *Viruses*

Many viruses insert their own DNA into the host's DNA document, where it gets copied and becomes part of the genome. More than half of the DNA in the human genome originated as invading viruses. These viruses can carry useful genes from other species into the infected animal or plant. Many human genes originated in other species by this mechanism.

### *Mutations*

With billions of letters in a DNA document, there are always a few copying errors, called mutations. A mutated gene has a changed meaning. It may specify a slightly different protein, or it may switch another gene on or off at a slightly different time. These changes are usually harmful, but tolerable, especially when we have two copies of a gene and one is unchanged. Occasionally a mutation is either deadly or beneficial. According to the theory of evolution, all of our genes originated, in the distant past, as mutations of other genes, a completely random process. The critics of genetic engineering mean to leave you with the impression that nature has evolved a finely tuned but fragile system of inheritance but that genetic engineers have no good idea what they are doing. But actually, the genetic engineers make only small well controlled modifications to a genome, whereas nature often makes large and random changes. The likelihood of unexpected effects in transgenic technology is small. Unexpected changes in conventional breeding are virtually certain.

## *Glyphosate*

We earlier mentioned glyphosate, an environmentally benign herbicide. It is biodegradable, lasting approximately two days after use, and its molecules bind to soil so it does not wash into streams or enter groundwater. Even if some madman were to dump it into a waterway, it is 230 times less toxic than the herbicide it replaced. It has been found safe and approved for use in almost every country, including all the countries of the European Union. Yet the anti-transgenic crowd seeks to ban it in Europe using the pretext of uniformizing regulations from one European country to another. What reason is given for banning the safest of all herbicides? Two reasons. It is supposed to be 'implicated' in non-Hodgkins lymphoma, and it is alleged to harm beneficial insects. Examining the evidence for these two claims reveals how some opponents of GMOs will use anything, no matter how shaky, to achieve their purposes. Non-Hodgkins lymphoma patients were asked to recall what pesticides they had been exposed to in the past several years. A statistically insignificant number of them (four) mentioned glyphosate, not surprising since it is so widely used. Even though the investigators considered this association meaningless, it was enough for Greenpeace to demand a ban. The claim about harming insects is even weaker. Several species were exposed to glyphosate and the control group were not exposed. There was similar mortality in both groups.

But there was mortality. This is enough for Greenpeace to claim that glyphosate kills beneficial insects. Greenpeace ignores a follow-up study by the International Organization for Biological Control, which concluded that glyphosate was exceptionally safe. One might have expected the IOBC to be hoping that it would find something wrong with glyphosate, since its charter is to popularize biological controls. But IOBC's scientists are ethical. The real reason for banning glyphosate is that it is made by Monsanto and used as the herbicide for the herbicide-resistant transgenic soybeans.

## *Labelling Transgenic Food*

Once the opponents instigated a doubt about the safety of genetic engineered foods, and remember that there has not, in ten years, been even the slightest evidence of that, the next step was to agitate for labelling. Why, after all, shouldn't consumers have a choice? I agree people who want to avoid transgenic food should be able to make that choice. But labelling can take two forms. One could label food which is not transgenic, or one could label food which is transgenic. I can't imagine anyone objecting to labelling foods which are transgenic-free. But the anti-transgenic demand is adamant for the other choice. We can say that the difference is between those who would label the non-transgenic food with a smiley face and those who would label transgenic food with a skull and crossbones. Labelling sounds so reasonable. How could anyone oppose it if they didn't have something to hide? But labelling has a down side. It costs.

The cost is not in the ink to print the label. It is in keeping the products separate. We are talking now mostly about soybeans and corn. These grains are harvested by the hundreds of tons, shipped in railroad cars, stored in grain silos, sold in futures contracts. Approximately 70 per cent of processed foods in the US have some ingredient derived from biotechnology.

This includes most milk and cheese, sugar, soy products, corn and corn sweeteners, vegetable oils, etc. 63 per cent of US grown soybeans are now transgenic. Neither buyers nor sellers distinguish between transgenic and conventional soybeans. Your tofu is a mixture of both kinds. (In my local supermarket, there is only one brand of tofu available and the manufacturer has recently decided to use only organic soybeans, which are specifically labelled non-GMO. At the same time, the package size changed from sixteen to fifteen ounces and the price was raised by $.25.) To keep the two kinds separate, we would need, at a minimum, to have two separate distribution channels, two storage systems, two futures markets. In my opinion, the pressure for skull and crossbones labelling is really a pressure to increase the cost of the genetic engineered food. So far, the transgenic food has had only producer advantages — it is cheaper to produce. Take away its cost advantage and it is no better than the conventional foods.

Never mind that corn with the Bt trait, one third of the US crop, is grown without pesticides. Never mind that the transgenic soybeans prevent soil loss and global warming. The Economist magazine estimated that segregation of grains would add a 25 per cent premium to the price of some processed foods, like packaged cereals. This is probably an over-estimate, but we should remember that these costs will fall inordinately on people with low income. But if this were the only problem with labelling, the food industry giants would quickly adopt grain segregation and labelling. They are being forced into grain segregation anyway by the labelling regulations of nations that import US grain.

The industry is really concerned that the labels will make it easier for groups like Greenpeace to boycott their products. Stop the scare campaigns and the resistance to labelling will disappear overnight. If you doubt that demands for skull and crossbones labelling can be used to purposely create a disadvantage, imagine what your reaction would be if someone were to propose that products assembled by hispanic workers must be so labelled. The public demand for labelling would be justified by a "right to information", but just below the radar screen you would not be surprised to hear that hispanic workers might be illiterate, or might be illegal immigrants, or might be drug users. But you would recognize that the demand for labelling was meant to disadvantage hispanics. We have established a precedent in this country about labelling. The government does not mandate labelling without a very good reason. The main exception is when safety (skull and crossbones label) is involved. Once the government decides that, say, a certain chemical pesticide is safe, nobody can require the food grown with that pesticide to be labelled. When a segment of the consuming public wants a label (smiley face) about a trait that it cares about, the market provides such a label and it is reflected in the cost of the product.

For example, there are people who prefer to eat food grown with no chemical pesticides. They buy "organic" food. Everyone understands that the label "organic" means that the food was grown without pesticides. Everyone also knows that they have to pay more for the food labelled organic. Similarly, orthodox Jews have certain religious rules about what they eat. These include a requirement that cows and chickens must be slaughtered in a particular way. This is a requirement for the Kosher label. Jews do not expect all meat to have a label detailing how the animal was slaughtered. Even in Israel, the people who want the label pay for the privilege, and kosher meat is often quite a bit more expensive that ordinary meat.

### *Environmental Scare Stories*

We mentioned earlier a concern that some genetic engineered crops could cross breed with wild relatives. To the pseudo-environmental groups (those who like to call themselves green even though much of what they advocate would harm the environment), this absolutely rules out any permissible use. Their argument goes as follows: "We don't have any idea what plants could cross breed with what other plants, and we don't have any idea what effect the unusual genes would have in the wild plants. Therefore we should take no chances." In fact, hybridization between different species is extremely uncommon and it has never been observed between distantly related species. A tomato won't cross with a potato, even though they are both members of the nightshade family.

Yet the GE opponents would have us believe that just about any plant can hybridize with any other. And they have quotes from PhDs to prove it.

Actually there is exactly one controlled experiment showing a hybridization between a genetic engineered crop and a wild species. The crop was a transgenic canola (oilseed), with that same herbicide tolerance gene we have encountered before. Here is what PhD Jeremy Bartlett, of the John Innes Plant Research Centre of Norfolk, England, wrote to the Manchester Guardian. He said we don't know what plants will hybridize with what other plants. He said that there is a documented example of transgenic canola hybridizing with a wild mustard. He went on in the same letter to talk about the possibility of gene transfer to soil organisms. Dr. Bartlett is a PhD in biology. He surely knows that there is little basis for speculating that plants will pass genes on to soil organisms. But let's give him the benefit of the doubt where a speculation is concerned. Still he must have known that the documented case he was referring to was observed at a test plot of the John Innes Research Centre, his own institution.

The transgenic canola was planted in the centre of the plot and various other species were planted at various distances to measure the rates of hybridization. There was only one case observed. The canola hybridized with the wild mustard. Jeremy Bartlett means for you to think that if an oilseed can hybridize with a wild mustard, then anything can hybridize with anything. Fortunately, the John Innes Centre puts its research reports on the world wide web, so I read the actual report. Guess what? Canola is a hybrid itself, a cross between Brassica rapa and Brassica napus, two closely related wild plants in the same genus. Brassica rapa is wild mustard! So we don't have evidence for a transgenic crop hybridizing with just about anything else imaginable, beyond the capability of man to anticipate. We have evidence of a plant crossbreeding with its closest relative, placed in the test plot because it was so likely to crossbreed, and it was the only case observed.

When the biotechnology opponents write or talk about this, they always say wild mustard and canola, never Brassica rapa and napus. Dr. Bartlett means to play with our minds, to mislead us. Hybridization with wild plants is a concern, but it is a managable concern.

## TRANSGENIC ORGANISM

A transgenic organism is an organism which has been modified with genetic material from another species. The genetic modification is accomplished by inserting DNA into an embryo with the assistance of a virus, a plasmid, or a gene gun. The embryo is allowed to develop, and the mature organism will express the DNA which has been inserted into its genome. Transgenic organisms can also pass the modification on to future generations by breeding with other members of the same species. Organisms may be genetically modified for any number of reasons. Genetic modification may be performed to make organisms more vigourous, to add resistance to specific threats, or for the goal of expressing a particular trait. It can be used for everything from adding nutrition to staple crops for the benefit of people in the developing world to making fish glow so that they can be sold as novelty pets. While species cannot interbreed, as a general rule, DNA from one species can express in another. This is because DNA codes for the same thing no matter where it is, so if a researcher can determine which part of an organism's genome codes for a particular trait and the trait is compatible with the organism being modified, the DNA can be successfully inserted to cause the trait to be expressed in a different species.

## *Rice with Vitamin A*

Rice does not contain very much vitamin A. In the poorer parts of Asia, where rice is almost the only food of the rural population, a vitamin A deficiency is common, leading to early blindness. Now Drs. Ingo Potrykus and Peter Beyer, two genetic engineers, have transferred the genes for vitamin A from other species into rice, creating a strain of rice which is rich in vitamin A — the amount of rice in a typical third world diet could provide about fifteen percent of the recommended daily allowance of vitamin A, sufficient to prevent blindness. Now that a few plants with this trait have been created, they are being cross bred with other varieties of rice using conventional breeding techniques, as has been done for centuries.

Such cross breeding could further increase the vitamin A content. The development of rice with vitamin A was carried out at the Swiss Federal Institute of Technology, making free use of patented technology and of the earlier research which had established the basic facts about how plants synthesize vitamins. The corporations holding the various patents all agreed to cost-free use of their patents as long as the rice was to be provided free to poor third-world farmers. The new rice strain was then turned over to the International Rice Research Institute, a non-profit organization based in the Philippines, where it will be evaluated for its adaptability to various growing conditions, food safety, and environmental impacts, etc.

The IRRI preserves thousands of varieties of rice with different genetic characteristics, so the new strain can be cross bred to produce varieties suitable for almost any locality. The result is that rural Asians can soon expect to retain normal eyesight. Genetic engineers also intend to produce a rice variety rich in iron, because iron-deficiency anemia is a common problem in the same rural populations. But this is a more difficult problem than increasing rice's vitamin A content. Rice contains a substance called phytate. Phytate prevents the body from absorbing iron, so it does little good to breed for increased iron content, and the rice plant cannot reproduce without adequate phytate in the grains. Dr. Potrykus hopes to be able to find a gene coding for a protein that will break down phytate when the rice is cooked.

## *No-Till Agriculture*

The world's biggest environmental problem is loss of topsoil to wind and drainage. The US experienced its dust bowl during the early part of the 20th century as a result of ploughing up the prairie. The problem is much worse in tropical soils, which may have a thin, inches thin, layer of topsoil above a type of soil which, when ploughed, turns into a non-porous concrete-like substance. One field, one crop, once. The problem, both in the prairie and in the tropics, is deep ploughing, which kills weeds which would otherwise crowd out the desired crop. The solution is called low-till agriculture.

The soil is broken up but not deeply ploughed. Weeds are killed instead by herbicides. A herbicide of choice should be cheap, quickly biodegradable and non-toxic. An excellent choice is a chemical called glyphosate, except that glyphosate kills the crops as well as the weeds. So genetic engineers found a gene which lets plants tolerate glyphosate, and transferred it into soybeans.

Today, 63 per cent of the soybeans grown in the US are glyphosate tolerant, allowing soil saving no-till agriculture on half the US soybean acreage. There is another advantage to no-till agriculture. There are lots of plant residues beneath the ground, both root systems and humus transported by earthworms. Ploughing brings this material to the surface, where it can oxidize.

Carbon dioxide is created, a greenhouse gas. So transgenic soybeans are a positive factor in postponing global warming. Approximately four tons of carbon dioxide are retained in the soil per acre per year. This saving is applicable to the accounting of $CO_2$ reductions in the Kyoto Treaty on Climate. In fact, any technology which reduces the need to plough, spray, or till crops will reduce carbon dioxide emission. Consider a tractor pulling a ten foot wide harrow over a square mile of agricultural land. Simple arithmetic shows that the tractor will travel 528 miles, all the while burning gasoline. Perhaps you are thinking that even if ploughing has disadvantages, herbicides don't sound very good either. The very word means "plant killer". But that is not the choice. Traditional soybeans are also grown using herbicides, most of which are far more toxic than glyphosate. On the average, the transgenic soybeans actually use 30 per cent less total herbicide than conventional soybeans. So environmentally, this is a no-brainer.

### *Witchweed Control*

Farmers in east Africa are plagued by a devastating parasitic weed called Striga, or witchweed. Farmers are used to dealing with weeds that grow in the soil alongside the crop and compete for nutrients. From time immemorial, they have dealt with those weeds by pulling them up by hand.

Less neighbour intensive methods like spraying and ploughing are now common. But none of these methods work for the witchweed. Striga attacks plants directly, underground, even before the weed has emerged above the soil surface. It sucks nutrients from the seeds and the roots of the crop. In some parts of Africa, the striga parasite destroys as much as 80 per cent of the crop yield.

But now that a herbicide resistance trait can be transferred to a crop, scientists in Israel and Kenya, working together, have demonstrated a new strategy for striga control. Before planting the crop, they soak its seeds in a herbicide. The seeds are unharmed, but they become poisonous to the striga parasite. The seed germinates and sprouts without interference. By the time the crop is harvested, the herbicide has decomposed and disappeared. Their demonstration used herbicide resistant transgenic corn. The same strategy would probably work with Africa's other important grains, sorghum and millet. Soaking seeds would use far less herbicide than spraying it on the ground, and the complex spraying apparatus would not be needed. This is a significant consideration in Africa, where so many farmers are too poor to own expensive equipment.

### *Saving the Banana*

Wild bananas have seeds. They reproduce sexually, like beans and oak trees. But they aren't easy to eat. Bananas grown on plantations have no seeds. They are reproduced by taking cuttings from older banana plants. Cultivated bananas are seedless because they have three of each type of chromosome instead of the normal two of each type. Such plants are called "triploid". They are always sterile. Genetic triploid freaks arise from time to time in nature, but modern breeders can also use chemicals or electric shocks to create triploid mutant cells. Bananas have been cultivated for many thousands of years and there are about three hundred different banana varieties. Each variety was developed by crossbreeding wild bananas. Whenever a promising variety was been produced, the breeder caused it to be triploid, hence seedless.

That plant was cloned by propagating cuttings, and it became the parent of its variety. All bananas plants of the same variety are genetically identical, like identical twins. Of the three hundred varieties, only one single variety completely dominates international trade. It is called Cavendish. It is possible that you have never seen a banana other than a Cavendish banana. Certain kinds of fungus can infect and kill banana plants. In many parts of the world, Cavendish banana plants are being attacked by a fungus called black Sigatoka. Since wild bananas can reproduce sexually, they are not all identical and some wild bananas can resist black Sigatoka. If the black Sigatoka fungus is present in a region, the resistant banana types become prevalent. But the Cavendish banana plants have no resistance.

They are all identical so they all die. To grow bananas commercially, growers must spray their plants with fungicides. Year after year, the black Sigatoka fungus has been evolving resistance to these fungicides, so growers have to spray more and more fungicide each year. Approximately one third of the cost of raising a banana is the cost of spraying it with fungicides, and it gets more and more costly each year. A form of black Sigatoka banana disease, now spreading around the world, can tolerate all known fungicides. Soon it will attack bananas in Central America and the Carribean islands, the heartland of banana culture. The Cavendish banana will become virtually extinct. Agronomists estimate that this will happen within ten years.

This is not a fairy tale. It has happened before. Forty years ago, the most popular variety of banana was one called Gros Michel. But Gros Michel was susceptible to a fungus called "race 1 Panama disease". Now it is gone. Cavendish bananas, which are not susceptible to race 1 Panama disease, replaced them. (A related fungus, race 4 Panama disease, can infect Cavendish banana plants. At present it is only found in Malaysia and some nearby countries.) In turn, some other variety of banana could replace the Cavendish. It would look different and taste different but it would still be a banana.

There is only one practical way to save the Cavendish banana. It must be given some combination of genes from wild bananas which are not susceptible to the fungus. But this can't be done by any natural technique. Cross-breeding can create other banana varieties because wild bananas reproduce sexually. But Cavendish bananas reproduce only by cloning. Conventional breeding would have to rely on rare mutated Cavendish banana plants which can produce seeds and which can therefore be crossbred, in theory.

Even the mutated plants produce only a tiny number of viable seeds, as few as two or three seeds in a hundred pounds of bananas. No banana breeding experts think that they can breed fungus resistance into a Cavendish banana variety in only ten years. Biotechnology can rescue the Cavendish banana. Scientists in Belgium have used genetic engineering to transfer some fungal resistance genes from wild bananas. The transformed Cavendish plants are not susceptible to the fungus and they can then be reproduced into nursery stock by the usual method of taking cuttings.

### *The Eggplant in Winter*

The edible part of an eggplant is formed from the ovary of its flower. In this way, it is like the edible flesh of an apple, a pepper or grape. When we eat these fruits, we discard the seeds. But the plants only transform their ovaries into fruits when they start to produce seeds, although in the case of an eggplant, its seeds are so tiny that we ignore them. Eggplants will only set seeds in warm weather, so to grow them in the winter in an unheated greenhouse, the grower must use a chemical to trick the plant into beginning fruit development without setting seed. Such fruits do not grow very large or very fast under these conditions. So eggplants are expensive in the winter.

But now scientists in Italy have transferred two genes into a variety of eggplant, which not only allows the plant to set fruit in cool greenhouse conditions without chemicals, but also increases productivity of the same plant in either hot or cold weather. The eggplant variety that the Italian scientists created is seedless. One of the two transferred genes is a switching gene which is turned on only in the ovary part of a flower. That gene turns on the other transferred gene, which makes a protein involved in synthesizing a growth hormone. The growth hormone makes the ovary grow into the fruit, just as it would have done in a traditional eggplant making seeds. Neither gene requires either seed setting or warm weather. Where does one get seeds to produce large numbers of seedless eggplants? The transformed plants produce pollen, so they can be crossed with traditional eggplant varieties and the hybrid produced by that crossing has the seedless and self-starting property. The scientists report productivity increases of 37 per cent for the new variety, and they believe that the seedless type would be more marketable.

## TRANSGENIC CROPS

The application of recombinant DNA technology has revolutionized plant breeding in recent years. Genetic transformation in plant breeding greatly increases the gene pool and broadens the scope of genetic changes that modern breeders may draw upon. The first generation of genetically modified (GM) crop plants hold great promise in carrying new or improved agronomical traits that require less intensive farming methods or which have improved quality traits. These GM crop plants have been produced to improve farming production systems, aiming, for instance, to reduce negative impacts on the environment, such as fertilizers, herbicides and pesticides.

In general they are of little direct interest to the consumer. However, the genetic manipulation of the primary and secondary metabolism of plants offers distinct possibilities in the development of nutritionally improved products, raw materials with added pharmaceutical value, and crops with real health benefits for the consumer. These are second-generation GM plants. Recently, several prototypes in molecular plant breeding with novel traits for (non-) food and industrial applications have been introduced into the market, and many more GM crops have the prospect of commercial exploitation in the near future.

The safety and wholesomeness of these GM crop plants raises questions beyond those posed by conventional foods. In this context, traditional plant breeding of most food crops has conformed to the standard 'generally recognized as safe' (GRAS). In some instances safety is assumed to be assured during performance and quality testing by breeders (*e.g.* glucosinolates and erucic acid in rape seed). However, the process is largely focused on the comparison of agronomic and phenotypic characteristics, such as colour, maturity, yield or disease resistance. Products with an unusual taste or that have harmful effects following ingestion or skin contact have normally been rejected from the breeding programme. However, the safety of conventional plant varieties has usually been presumed from the assumption that prudent consumers will avoid those species once it is known that they cannot consume them without adverse health effects. When the commercial production of GM plants started, it sparked a debate centred on issues associated with concerns on environmental, human and animal safety matters. Many people questioned whether the 'simple' GRAS approach, based on historical and empirical experiences, is sufficient to guarantee food safety of the GM crop plants.

It is also asked whether the risks of chronic exposure of humans and animals to GM crops are adequately covered, and whether large-scale breeding of GM plants will have detrimental effects on ecosystems.

## Food safety issues

Although the variety registration of existing crop plants has not resulted in adverse effects in humans, the conventional assessments carried out by plant breeders was not considered to be sufficient to ensure food safety of GM crop plants. As a result, Europe issued a regulation on novel foods and novel food ingredients that sets the legal framework for the market introduction of genetically modified organisms (GMOs). This came into force on 16 May 1997.

The Novel Foods Regulation establishes a system of mandatory pre-market approvals of novel foodstuffs, including GM plants. The accompanying EU guideline 97/618/EC gave a clear indication of the types of data that are needed to form the basis of an assessment. In particular, four central safety issues have been set forth related to food and feed safety:

- The nutritional and toxicological consequences of inserted gene products;
- The potential of pleiotropic (unintended) effects in the host organism due to the insertion event;
- The allergenicity of expressed proteins and novel foodstuffs;
- The potential of gene transfer to human and animal gut flora.

This means that in transgenic insect-resistant Bt tomatoes encoding the C-terminal truncated Bt2 gene derived from a *Bacillus thuringiensis* (Bt) strain, *IAb5*, the novel protein has to be evaluated on its own merit, while the safety and wholesomeness of the 'remaining' transgenic tomato is assessed separately. Whereas, for example, in antisense RNA exogalactanase tomatoes, which have improved rheologic characteristics due to downregulation of the endogenous exogalactanase activity, the safety assessment is primarily focused on the 'remaining' novel fruit. In general, there is basic agreement on the safety issues to be addressed. One particular area of concern is the possibility of unexpected or unintended metabolic perturbations due to genetic modification that may alter, for instance, levels of nutrients and health-influencing components. The use of rDNA techniques does not necessarily result in fundamental changes in crop plants compared to the food produced by conventional breeding methods. However, it should be emphasized that a uniform international agreement for the evaluation of GMOs is urgently needed.

For example, an agreement on how to establish in practical terms the similarities or differences between the 'remaining' GM crop plant and the traditionally bred plant seems to be much more difficult to achieve. This chapter is devoted to the description of advanced strategies for the evaluation of potential unintended effects in GM crop plants. Pleiotropic (unintended) alterations in agronomic traits or composition may arise from insertion mutagenesis or as a result of metabolic effects of the novel gene product(s). The challenge is to gain the ability to discriminate between metabolic alterations due to somaclonal variations, unintended effects or natural diversity. Thereto, emphasis is given to a platform of technologies designed to identify changes in the molecular machinery of crop plants. As is the case for detection methods for GM crops and derived products, there are needs and criteria for sampling strategies, statistical models and manufacturing standard reference matehers recently.

## VIRUS RESISTANT CROPS

Some viruses infect people or animals and other viruses infect plants. Plant viruses reduce the productivity of annual crops and can kill fruit trees. Some plant viruses are spread by insects. Plants can be protected from those viruses by using insecticides or other pest management methods. There is essentially nothing else that a farmer can do to protect his crop from virus damage, except to grow a different crop. But genetic engineering a plant to protect it from a particular kind of virus is quite easy.

A gene from the virus which encodes a protein in the virus' outer coat is copied into the plant's DNA. The plant then makes the coat protein, which is harmless, but which stimulates the plant's natural defences. Virus resistance traits have been introduced into many crops, including squashes, tomatoes, potatoes, tobacco and, perhaps most dramatically papaya. Recently, cultivated Hawaiian papayas were hit by a devastating virus which essentially eradicated the commercial variety. Only the virus resistant transgenic papayas survived. If you like papaya, you can only buy the transgenic variety. Nobody can grow any other kind.

### *The Potato Famine*

In 1840s Ireland, the potato crop was devastated by a late blight fungus (Phytophthora infestans) and Irish people starved en masse. That fungus could reappear at any time in any place and wipe out a potato crop. Some varieties of potato have previously had some resistance to late blight fungus, but now a fungal strain has appeared in Russia that destroys those previously resistant varieties.

This year (2001) a similar fungus appeared in potato fields in Prince Edward Island and 630 million pounds of potatoes, the island's principal crop, had to be destroyed. But very recently, scientists were able to transfer a gene from alfalfa to a potato plant and the resulting potato plant is able to resist the fungus and thrive. Potatoes also rot. A principal cause of potato rot is the bacterium Erwina carotovora, which has been called the flesh eating bacteria of the plant kingdom. Now a gene that confers resistance to E. carotovora has been coupled to a control gene that turns on when a plant has been wounded, and this construct has been transferred to experimental potatoes. As the researchers hoped, the modified potatoes, when punctured by a toothpick and exposed to E. carotovora, had almost twenty times less rot than unmodified potatoes.

### *Sentinel Crops*

A recent innovation is a plant intended not for food but for quality control. It contains a gene derived from a luminescent jellyfish, but in all other ways it is identical to the food crop it is planted alongside. When these sentinel plants experience a lack of water, they literally glow in the dark. The farmer then knows that his crop must be watered or whether irrigation can be postponed. In the western U.S. water is scarce. Agriculture is the biggest user of water. Wasting water is intolerable. For example, so much water is taken from the Colorado River for irrigation that the river flows into Mexico a mere trickle, and it never gets to the sea at all. So this is yet another way that transgenic crops can benefit the environment.

### *Building with Silk*

Silk is composed of two proteins, fibroin and sericin. The gene for fibroin has been transferred from a silkworm to a goat, and is expressed as a component of its milk. Soon we may also expect sericin to be transferred. It still remains to be seen whether technology can be developed to spin these proteins into a fibre. That has already been accomplished with the kind of silk spiders use to make their webs. Genes for the two spider silk proteins were transferred to cells cultured from cow udders. Those cells then made the proteins. Happily, the spider silk can be spun by forcing a solution of its two protein components through a tiny nozzle. The proteins self-assemble into spider silk strands. The same genes have since been transferred to live goats and when those goats are old enough to produce milk, it should be possible to make large quantities of spider silk very cheaply. Silk is an extraordinarily strong material, stronger than steel. In the future we may be getting our strongest building material from a farm instead of from a mine.

### *Safer Meat*

Escherichia coli are friendly bacteria that live in our intestines and contribute to our health. But there is one strain of E. Coli (designated as O157:H7) that can make us sick, even kill us. We can get it from inadequately cooked meat. The E. coli infected meat comes from cattle with the virulent E. coli strain in their intestines.

A cow's digestive system is adapted to digesting hay and grasses. The food first goes into a pre-stomach called a rumen. That's why cows are called ruminants. In the rumen, microorganisms turn the indigestible cellulose into nutrients the cow can assimilate. The food is then passed to a true stomach, and finally gets to the cow's intestines, where E. coli can live.

Therefore to guarantee against the virulent strain thriving in the cow's intestine, one needs to get some sort of prophylactic agent into the intestine. Antibiotics won't do. They would kill the cow's normal intestinal bacteria, and besides, it isn't a good idea to overuse antibiotics. There are antibodies specific to the virulent strain of E. coli, but they would be destroyed by passage through the cow's stomach before reaching its intestines. Genetic engineers are working on a neat solution.

They are developing a transgenic animal feed which resists complete digestion in the stomach and delivers the antibody, specific to the virulent E. coli strain, into the intestines. There are over 60,000 cases of E. coli illness in the United States each year. There would be many more except for an extensive programme of meat inspection. Even this understates the problem, because meat, if contaminated, has to be destroyed.

E. coli infections were not such a serious problem when cattle were raised exclusively on grass and hay. On that diet, there isn't much digestion going on in the cow's intestine and the E. coli populations are comparatively low. But today's cattle spend the last weeks of their lives in feed lots, being fattened up on grain, which is digested in their intestines, leading to much higher populations of E. coli. So another way to solve the E. coli problem would be to raise leaner cattle and skip the feed lots.

### *Reduced Need for Fertilizers*

One of the ways that farmers get better yields is by providing their plants with sources of organically bound nitrogen and phosphorus. These can be provided either by applying chemical nitrates or phosphates, or by using manures or decaying vegetation as sources of the same nutrients. Nitrogen is the largest constituent of the atmosphere, about 80 per cent. It may seem paradoxical that unfertilized plants could suffer from a nitrogen deficiency while immersed in a sea of nitrogen gas, but it is just not available in the form they need. Of living things, only certain bacteria (and human chemists) have evolved the means to convert nitrogen from the atmosphere to a form useful to plants. But some plants, primarily legumes (peas and beans), have a symbiosis with these nitrogen fixing bacteria. The plants provide nodules on their roots that protect the nitrogen fixing bacteria, which then enrich the soil around those roots.

Not only does this permit the legumes to grow luxuriantly without nitrate fertilization, but it makes the soil fertile for other plants growing in the same soil later. The technique of crop rotation is one of the oldest techniques of agriculture. Scientists hope to be able to transfer the genes which direct the formation of the nodules to other crops. If this is successful, the need for fertilizers would be dramatically reduced. Unlike nitrogen, phosphorus is not a constituent of the atmosphere. There is no short-term likelihood that scientists will find a genetic engineering way to replace fertilizers that provide phosphates.

The best hope for phosphate replacement would be to breed or engineer plants that make more efficient use of the phosphate available to them. If it proves impossible to engineer plants for nitrogen fixation, there are still options which can let them use fertilizers more efficiently. An enzyme called glucine dehydrogenase is involved in utilization of fertilizers. The gene for glucine dehydrogenase is present in most crops, but it is expressed at low levels, because the control genes turn it off more than on. A genetic transformation of wheat which promoted increased synthesis of glucine dehydrogenase was 29 per cent more effective in utilizing the same amount of fertilizer as the unmodified variety. The increased efficiency can either be used to grow more crop on the same land, or to cut down on the need for fertilizer to grow the same amount of crop.

# 2

# Applications of Genetic Engineering of Food Crops

## APPLICATIONS OF GENETIC ENGINEERING

### "Pharming" and "plantibodies"

An increasingly viable option is the production of highly valuable enzymes by plants and animals. In addition to production of human proteins in these organisms, other valuable proteins that are currently produced by microorganisms could very well be produced by higher organisms instead. Animal and plant "bioreactors" in some respects may be superior to recombinant bacterial systems, because eukaryotes glycosylate proteins. Whereas the glycosylation pattern may be species-specific, appropriate glycosylation is often required for protein function. Production through these organismal systems may also be cheaper than cell fermentation techniques.

Two examples of production of human proteins in plants include the production of human serum albumin in transgenic tobacco and potato, and production of human insulin by tobacco. In both cases, the produced protein appears to be fully effective in humans. Unfortunately, however, one cannot raise his/her insulin level by eating transgenic tobacco leaves, as the protein in most cases will be broken down to amino acids before it reaches the blood stream.

Therefore, in these cases one cannot escape the practice of protein isolation and purification before transgenic leaves are converted into drugs. Also antibodies are being produced in plants. Initially, the antibody's light and heavy chains were produced in different plants. But a subsequent cross of these two varieties resulted in progeny carrying assembled and functional antibodies.

## Reversible male sterility in plants

As has been indicated earlier, heterozygous individuals often are healthier and stronger than homozygous ones. The only way to guarantee heterozygoiscity in plants is to make sure self-pollination cannot occur. For most crop plants it was very tedious or practically impossible to exclude selfing.

To exclude self-pollination, it would be good to introduce male sterility in plants: progeny from such plants are then expected to be 100percent heterozygous (assuming they were pollinated with pollen from an unrelated variety). To introduce male sterility, a promoter was identified that was turned on exclusively in tapetum cells (a tissue around the pollen sac that is essential for pollen production). This promoter then was linked up to a gene coding for a bacterial ribonuclease (named barnase). This ribonuclease selectively chops up ribonucleic acids. The promoter/ribonuclease construct was then introduced into plants (canola, tobacco, you name it).

Because the promoter allows expression only in tapetum cells, the gene construct disrupts only development of the tapetal tissue and its end product, pollen. Plants transformed with this construct were male-sterile but otherwise normal. Although male-sterile plants are valuable for hybrid seed production, they have limited value when it comes to crop production. Fertility must be restored to crops such as wheat, rice, and tomato, in which the seed or fruit is the harvested product. Fortunately, the ribonuclease is inhibited very much by a simple protein, named barstar. One can thus cross the male-sterile plant with a male-fertile variety in which the gene for barstar has been introduced, and the result is progeny with viable pollen and restored fertility. A closely related approach has been criticized as "terminator technology" as it is seen by its critics as a way for companies to protect and enforce their patents.

## Antisense RNA

Antisense RNA refers to nucleotide strands that are produced in a cell and that are complementary to a particular mRNA. Antisense RNA can be produced, by inverting the coding region of a gene with respect to its promoter. The antisense RNA can hybridize with its corresponding mRNA, making it double-stranded. The double-stranded mRNA no longer can be recognized by the protein-synthesizing machinery (the ribosomes), and thus expression of this mRNA is suppressed.

Also, in many systems double-stranded mRNA is very unstable and is broken down quickly. Thus, one can inactivate specific genes while not interfering with others. Antisense approaches already are used to protect plants from damage by plant viruses. Reversal of a gene from bean yellow mosaic virus (BYMV), and putting it into tobacco under a reasonably strong promoter, has led to a tobacco variety that is quite resistant to BYMV.

A similar approach is used to transfer viral resistance to other plants. This finding is of significance, in that currently no effective, environmentally friendly methods exist to control many plant viruses. Very related to this approach is the RNAi (RNA interference) approach. This is very useful for both agriculture and medicine, and the first examples of practical applications of this RNAi technology are appearing. As with any new technology, the initial pilot projects are sort of pedantic. However, more exciting application possibilities abound.

Obviously, RNAi technology provides an excellent approach for reverse genetics in eukaryotes. With this method one can turn off genes. Perhaps indicative of the large potential and relative ease of genetic engineering, developing countries (particularly China) are progressing rapidly in development and application of genetically engineered crops. Some have gone into commercial production well ahead of similar crops in the US. In China, tomatoes that have been engineered for improved virus resistance have been on the market since late 1992. There are two main reasons for the more rapid commercialization of bioengineered crops in the developing world:

- Less tight governmental approval mechanisms
- Hungrier populations.

While in developed countries the main value of biotechnological applications may be to reduce production costs, in the developing world a main factor is the production of more food. Indeed, genetic engineering applications seem to be pretty successful to cut down on pathogen-induced losses. Genetic modification of papaya plants (expression of the ringspot virus coat protein in the plant) protects very well against the very destructive ringspot virus.

## CELL RESPIRATION

As mentioned in the previous page on ATP, the process of *respiration* is split into 3 distinct areas that occur at different parts of the cell. Respiration involves the *oxidation* of foodstuff (*i.e.* glucose) in order to create ATP. Respiration can occur with or without oxygen, *aerobic* and *anaerobic* respiration respectively.

### Glycolysis

Glycolysis occurs in the *cytoplasm* of a cell where a 6 carbon glucose molecule (the broken down food that you ate earlier) is broken down by enzymes into a 3 carbon *pyruvic acid*. The execution of this process requires 2 ATP, and produces a net gain of 2 ATP.

The enzymes involved remove hydrogen from the glucose (oxidation) where they take these hydrogen atoms to the cytochrome system, explained soon. In anaerobic respiration, this is where the process ends, glucose is split into 2 molecules of pyruvic acid. When oxygen is present, pyruvic is broken down into other carbon compounds in the Kreb's Cycle. When it is not present, the pyruvic acid is broken down into lactic acid (or carbon dioxide and ethanol).

## The Kreb's Cycle

When oxygen is present, respiration can harness more ATP from a single unit of glucose. The *pyruvic acid* from the *glycolysis* stage diffuses into a cell organelle called a mitochondrion (pl. *mitochondria*). These mitochondria are sausage shaped structures that host a large surface area for the respiration to occur on. The pyruvic acid is then subject to more enzymes which break it down into a 2 carbon compound, as seen below. The diagram illustrates the *Kreb's cycle*, consisting of three main actions

- The carbon element is in an infinite cycle where the 2 carbon compound derived from pyruvic acid binds with the 4 carbon compound that is always present in the cycle.
- $CO_2$ is released, where the oxygen that is present in *aerobic respiration* combines with carbon from the carbon compounds which is released as $CO_2$. Hence the need for animals to breath out and expel this $CO_2$.
- Enzymes oxidize the carbon compounds and transport the hydrogen atoms to the cytochrome system.

## The Cytochrome System

The cytochrome system, also known as the hydrogen carrier system (or the *electron transport system*) are where the reduced hydrogen carriers transport hydrogen atoms from the glycolysis and Kreb's cycle stages. The cytochrome system is found in the many *cristae* of mitochondria, which are tiny stalked particles found on its outer layer. The system contains many 'hydrogen acceptors' which hydrogen can be added to. By following the path of a hydrogen atom, we can see how the cytochrome system works:

- Some *coenzymes* from earlier stages are transferred to the next coenzymes.
- B is then oxidised, therefore the coenzyme releases the hydrogen and energy is made available.

- The released hydrogen atom binds with 2 oxygen atoms (oxygen is available in aerobic respiration) which produces water, a by-product of respiration.

The diagram illustrates this flow of hydrogen within the cytochrome system and how energy is made available by the flow of these atoms. The green circles illustrate where energy is made available via oxidation.

Overall their is a gain of 38 ATP from one molecule of glucose in aerobic respiration. The food that we eat provides glucose required in respiration. In plants, energy is also acquired via respiration, but the mechanism of delivering glucose to the respiration process is a little different.

## GENETICS AND EVOLUTION

### Meiosis - The Genetics of Reproduction

The genetic information found in *DNA* is essential in creating all the characteristics of an organism.

This remains the case when passing genetic information to offspring, that can occur via a process called *meiosis* where four *haploid* cells are created from their *diploid* parent cell. For a species to survive, and genetic information to be preserved and passed on, reproduction must occur. This can be done by passing on the information found in the chromosomes via the gametes that are created in meiosis.

### Chromosome Complement

Humans are diploid creatures, meaning that each of the *chromosomes* in our body are paired up with another. Haploid cells possess only one set of a chromosome. A diploid human cell possesses 46 chromosomes and a gamete created by a human is haploid possesses 23 chromosomes. Tetraploid organisms possess more than 3 sets of a particular chromosome.

### Reproduction

Reproduction occurs in humans with the fusion of two haploid cells (*gametes*) that create a *zygote*. The nuclei of both these cells fuse, bringing together half the genetic information from the parents into one new cell, that is now genetically different from both its parents. This increases *genetic diversity*, as half of the genetic content from each of the parents brings about unique offspring, which possesses a unique *genome* presenting unique characteristics. Meiosis as a process can increase genetic variation in many ways, explained soon.

## The Process of Meiosis

The process of meiosis essentially involves two cycles of division, involving a gamete mother cell (diploid cell) dividing and then dividing again to form 4 haploid cells. These can be subdivided into four distinct phases which are a continuous process

### *1st Division*

- *Prophase- Homologous chromosomes* in the nucleus begin to pair up with one another and then split into chromatids (one half of a chromosome) where crossing over can occur. Crossing offer can increase genetic variation.
- *Metaphase-* Chromosomes line up at the equator of the cell, where the sequence of the chromosomes lined up is at random, through *chance,* increasing genetic variation via independent assortment.
- *Anaphase-* The homologous chromosomes move to opposing poles from the equator
- *Telophase-* A new nuclei forms near each pole alongside its new chromosome compliment.

At this stage two haploid cells have been created from the original diploid cell of the parent.

### *2nd Division*

- *Prophase II-* The nuclear membrane disappears and the second meiotic division is initiated.
- *Metaphase II-* Pairs of chromatids line up at the equator
- *Anaphase II-* Each of these chromatid pairs move away from the equator to the poles via *spindle fibres*
- *Telophase II-* Four new haploid gametes are created that will fuse with the gametes of the opposite sex to create a zygote.

Overall, this process of meiosis creates gametes to pass genetic information from parents to offspring, continuing the family tree and the species as a whole.

## Independent Assortment and Crossing Over

The previous page investigates the process of meiosis, where 4 *haploid* gametes are created from the parent cell. Half the genetic information from a parent is present in these haploids, which fuse with gametes of the opposite sex to create a *zygote,* with a complete chromosome compliment that will create offspring after prolonged growth.

The process of *meiosis* increases *genetic diversity* in a species. The sex organs which produce the haploid gametes are the site of many occurrences where genetic information is exchanged or manipulated.

## Independent Assortment of Chromosomes

*Alleles* for a particular *phenotype* determine what characteristic an organism will express, as with the following example where

- Chromosome 1 contains an allele for blonde hair
- Chromosome 2 contains an allele for brown hair
- Chromosome 3 contains an allele for blue eyes
- Chromosome 4 contains an allele for brown eyes

The top assortment to the left produces 2 blonde hair/blue eyes gametes while the below produces 2 brown hair/brown eyes gametes. The top assortment on the right produces 2 blonde hair/brown eyes gametes while the below produces 2 brown hair/blue eyes gametes.

The above indicates that even though the two homologous chromosomes contain the same genetic information, the assortment of the chromosomes (the order they lie in) can determine what genetic information is present in each of the 4 gametes produced. With 23 chromosomes in a human gamete, their are $2^{23}$ combinations (8388608 combinations)

## Crossing Over

During meiosis, when *homologous chromosomes* are paired together, there are points along the chromosomes that make contact with the other pair. This point of contact is deemed the *chiasmata,* and can allow the exchange of genetic information between chromosomes. This further increases *genetic variation.*

## Crossing Over and Genetic Diversity

Gregor Mendel, an Austrian monk, is most famous in this field for his study of the *phenotype* of pea plants, including the shape of the peas on the pea plants.

## Gregor Mendel's Work

Mendel's goal was to have a firm scientific basis on the relationship of genetic information passed on from parents to offspring. In light of this he focused on how plant offspring acquired the phenotype of their seeds. The plants that were used in the experiment had to be *true breeding, i.e.* those plants with round seeds must have had parents with round seeds, who in turn had parents producing round seeds etc. This is done to increase the accuracy of results. After successfully producing two generations from these true breeding plants, the following was evident

- The first generation of plants produced all had a round seed phenotype.
- When these first generation plants were crossed, a ratio of 3 round seeds averaged every 1 wrinkled seed.
- The ratio of 3:1 was not exact, though this is because of the randomness of the processes that are executed to produce these plants. Independent assortment is completely random, as are mutations, therefore variable results occur producing a sampling error.
- Due to the scale of the experiment done by Gregor Mendel, the sampling error was smaller than that of a smaller scale experiment.

Mendel successfully hypothesised that the reason for this trend in phenotypes from generation to generation was down to the fact that genetic information was being passed on from their parents. The fact that round seeds appeared more frequently than wrinkled seeds is due to round seeds being the dominant phenotype, which when present effectively 'masks' the phenotype of the recessive (wrinkled seed) gene.

## Dominant and Recessive Alleles

- All plant seeds produced in the first generation were round
- 3 out of every 4 plant seeds produced in the second generation were round.

The parents, one possessing wrinkled seeds the other possessing round were crossed together, for some reason in the first and second generation the presence of the round seed gene in offspring superceded the presence of the wrinkled seed. This is called dominance.

## Mendel's Law & Mendelian Genetics

### *Mendel's First Law*

*"The alleles of a gene exist in pairs but when gametes are formed, the members of each pair pass into different gametes. Thus each gamete contains only one allele of each gene."*

## Incomplete Dominance

When a particular gene possesses both *dominant* and recessive alleles, it is possible for *incomplete dominance* to occur, where the organism at hand expresses a phenotype morphed by the expression of both the dominant and recessive alleles. In essence, *heterozygous* (possessing opposing alleles Rr) organisms derived from *homozygous* (possessing the same alleles RR or rr) are created, they possess a phenotype different to that of both their parents.

## Multiple Alleles

Diploid organisms naturally have a maximum of 2 alleles for each gene expressing a particular characteristic, one deriving from each parent. In some cases, however, more than two types of allele can code for a particular characteristic, as is the case of genetic coding for blood type in humans. Their are up to 6 possible genotypes that code for the four blood groups, A, B, AB and O.

## Example of a Cross

The following dihybrid cross involves two true breeding pea plants, where two factors are looked at, the shape of the seed and the colour of the seed.

## Mendelian Genetics

The past few pages have elaborated on the work of Gregor Mendel and how his work has paved the way to predicting the characteristics of offspring. However, a degree of randomness is involved, when involving factors such as independent assortment during meiosis and the possibility of genetic mutations.

## Chromosome Mutations

It is natures intention that the exact genetic information from both parents will be seen in the offspring's *DNA* in the the critical stages of fertilisation. However, it is possible for this genetic information to mutate, which in most cases, can result in fatal or negative consequencies in the outcome of the new ogranism.

### *Non-Disjunction and Down's Syndrome*

One well known example of mutation is *non-disjunction*. Non-disjunction is when the spindle fibres fail to seperate during meiosis, resulting in gametes with one extra chromosome and other gametes lacking a chromosome. If this non-disjunction occurs in chromosome 21 of a human egg cell, a condition called *Down's syndrome* occurs. This is because their cells possess 47 chromosomes as opposed to the normal chromosome compliment in humans of 46. The fundamental structure of a chromosome is subject to mutation, which will most likely occur during crossing over at meiosis. There are a number of ways in which the chromosome structure can change, as indicated below, which will detrimentally change the genotype and phenotype of the organism. However, if the chromosome mutation effects an essential part of DNA, it is possible that the mutation will abort the offspring before it has the chance of being born. The following indicates types of chromosome mutation where whole genes are moved:

### *Deletion of a Gene*

As the name implies, genes of a chromosome are permanently lost as they become unattached to the centromere and are lost forever

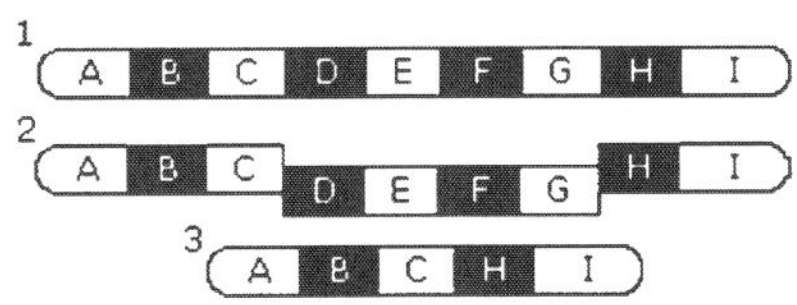


- Normal chromosome before mutation.
- Genes not attached to centromere become loose and lost forever.
- New chromosome lacks certain genes which may prove fatal depending.

### *Duplication of Genes*

In this mutation, the mutants genes are displayed twice on the same chromosome due to duplication of these genes. This can prove to be an advantageous mutation as no genetic information is lost or altered and new genes are gained

- Normal chromosome before mutation
- Genes from the homologous chromosome are copied and inserted into the genetic sequence
- New chromosome possesses all its initial genes plus a duplicated one, which is usually harmless

## Genetic Mutations

This page continues from the previous page investigating genetic mutations.

### *Inversion of Genes*

*This is where the order of a particular order of genes are reversed as seen below:*

- Normal chromosome un-altered
- The connection between genes break and the sequence of these genes are reversed

- The new sequence may not be viable to produce an organism, depending on which genes are reversed. Advantageous characteristics from this mutation are also possible

### *Translocation of Genes*

This is where information from one of two homologous chromosomes breaks and binds to the other. Usually this sort of mutation is lethal

- An un-altered pair of homologous chromosomes
- Translocation of genes has resulted in some genes from one of the chromosomes attaching to the opposing chromosome

### *Alteration of a DNA Sequence*

Mutation have investigated changes at the chromosome level. The sequence of nucleotides on a DNA sequence are also susceptible to mutation.

- Deletion: Here, certain nucleotides are deleted, which affects the coding of proteins that use this DNA sequence. A gene coded for alanine, with a genetic sequence of C-G-G, and the cytosine nucleotide was deleted, then the alanine amino acid would not be able to be created, and any other amino acids that are supposed to be coded from this DNA sequence will also be unable to be produced because each successive nucleotide after the deleted nucleotide will be out of place.
- Insertion: Similar to the effects of deletion, where a nucleotide is inserted into a genetic sequence and therefore alters the chain thereafter. This alteration of a nucleotide sequence is known as frameshift
- Inversion: Where a particular nucleotide sequence is reversed, and is not as serious as the above mutations. This is because the nucleotides that have been reversed in order only affect a small portion of the sequence at large
- Substitution: A certain nucleotide is replaced with another, which will affect any amino acid to be synthesised from this sequence due to this change. If the gene is essential, *i.e.* for the coding of haemoglobin then the effects are serious, and organisms in this instance suffer from a condition called sickle cell anaemia.

All of the genetic mutations looked at through the last 2 pages more or less have a negative impact and are undesired, however, in some cases they can prove advantageous.

Genetic mutations increase genetic diversity and therefore have an important part to play. They are also the reason many people inherit diseases.

## PLANT CELL DEFENCE

### *Hydrogen Peroxide*

Plants release *hydrogen peroxide* in response to the presence of a fungal invasion, which attacks by piercing the cell wall of a plant and breaking it down. This hydrogen peroxide (chemical symbol $H_2O_2$) is a double edged sword in its defence against the antigen.

*One Way*: Hydrogen peroxide stops the breakdown of the cell wall. Certain pathogens will use pectinase, a digestive enzyme, to break down the cell wall barrier and invade the plant. The *pectinase* released by the fungus must be stopped. $H_2O_2$ is involved in halting the action of this pectinase:

- The $H_2O_2$ is created and moves to the cell wall - the site of the invasion.
- It reacts in contact with and enzyme called *peroxidase,* which promotes the breakdown of *pectinase.*
- The foreign chemical is rendered useless.
- Threat of the cell wall being compromised is removed.

Another Way: Some of the $H_2O_2$ triggers the creation of phytoalexins

Phytoalexins are similar to the antiviral proteins previously mentioned as a secondary line of defence. Phytoalexins are a family of hormones that inhibit protein synthesis and thus "shut up shop" in the event of a pathogenic attack by halting the protein production process in our cells.

- Chemicals released by the fungus that are being used by it in its attack trigger a chemical response in the plasma membrane that makes the plant aware of the pathogens presence.
- Hydrogen peroxide from the plasma membrane triggers a chemical response to inform the nucleus of the infected cells of the current situation.
- mRNA from the nucleus is transported to ribosomes, as described in the protein synthesis section, where phytoalexins are to be produced (essentially protein synthesis coding for phytoalexins).
- The phytoalexins then take on a role similar to that of antiviral proteins, where the presence of a phytoalexin in a cell inhibits protein synthesis and therefore preventing growth of the foreign agent by removing all possible avenues of invasion for the pathogen, thus eliminating the threat.

### Barriers Used by Plants in Defence

*Lignin* is a strong type of molecule that provides plants with a defensive structure similar to that of fibrous proteins. It acts as a barrier and can be found in wood and is characteristically found in plants that have recently endured pathogen attack.

*Callose* seals off sieve plates in the plant, effectively shutting off the transport of molecules around the organism. This is done to minimise the chance of the plant transporting infectious material around its own self, and halting the movement of materials that could be used by the pathogen in aid of replicating itself. Ethylene promotes leaf abscission, and is done to sever the plant of dead or dying plant matter. This is done to prevent the spread of infected material, therefore sacrificing infected sections of plant is more economical than taking the risk of the infection spreading.

*Galls* and *tannins* are created by the plant to encapsulate foreign agents found within the plant. A gall is an instance where an infected cell becomes inflamed that contains tannins. These tannins play a protective role by segregating the foreign agent and its chemicals from the rest of the plant. All of the four previous pages have illustrated means of self defence against *pathogens* (fungi, viruses and bacteria).

## Plant transformation

Genetic engineering of plants is much easier than that of animals. There are several reasons for this:

- There is a natural transformation system for plants (the bacterium *Agrobacterium tumefaciens*),
- Plant tissue can redifferentiate (a transformed piece of leaf may be regenerated to a whole plant), and
- Plant transformation and regeneration are relatively easy for a variety of plants.

The soil bacterium *Agrobacterium tumefaciens* ("tumefaciens" meaning tumor-making) can infect wounded plant tissue, transferring a large plasmid, the Ti plasmid, to the plant cell. Part of the Ti (tumor-inducing) plasmid apparently randomly integrates into the chromosome of the plant. The integrated part of the plasmid contains genes for the synthesis of:

- Food for the bacterium.
- *Plant hormones*: Genes from the Ti plasmid that are integrated in the plant chromosome are expressed at high levels in the plant.

Overproduction of the plant hormones leads to continuous growth of the transformed cells, causing plant tumors. Rapid, cancerous growth of the transformed plant tissue obviously is advantageous to the bacterium: more food gets produced.

The Ti plasmid has been genetically modified ("disarmed") by deleting the genes involved in the production of bacterial food and of plant hormones, and inserting a gene that can be used as a selectable marker. Selectable marker genes generally are coding for proteins involved in breakdown of antibiotics, such as kanamycin. Any gene of interest can be inserted into the Ti plasmid as well. In principle, one can thus transform any plant tissue, and select transformants by screening for antibiotic resistance.

However, unfortunately, there are some complications:

- It has proven difficult to transform some monocots (grasses, etc.) by *Agrobacterium,* and
- Regeneration of plants from tissue culture or leaf discs is not always possible.

A number of genetically engineered plant varieties have been developed. Traits that have been introduced by transformation include herbicide resistance, increased virus tolerance, or decreased sensitivity to insect or pathogen attack. Traditionally, most of such genetically engineered plants were tobacco, petunia, or similar species with a relatively limited agricultural application.

However, during the past decade it now has become possible to transform major staples such as corn and rice and to regenerate them to a fertile plant. Increasingly, the transformation procedures used do not depend on *Agrobacterium tumefaciens*. Instead, DNA can be delivered into the cells by small, µm-sized tungsten or gold bullets coated with the DNA. The bullets are fired from a device that works similar to a shotgun. The modernized device uses a sudden change in pressure of He gas to propel the particles, but the principle of "shooting" the DNA into the cell remains the same. This DNA-delivery device is nicknamed "gene gun", and has been shown to work for DNA delivery into chloroplasts as well. Over the last several years, use of the "gene gun" has become a very common method to transform plants, and has been shown to be applicable to virtually all species investigated. Transformation of rice by this method is now routine.

This is a very important development as rice is the most important crop in the world in terms of the number of people critically dependent on it for a major part of their diet. Another method to get foreign genes into cereals is by electroporation: a jolt of electricity is used to puncture self-repairing holes in protoplasts (*i.e.,* the cell without the cell wall), and DNA can get in through these holes. However, it is often very difficult to regenerate fertile plants from protoplasts of cereals.

Nonetheless, significant advances in overcoming these practical difficulties have been made over the years. Now even transgenic trees have been created:

the gene for a coat protein of the plum pox virus has been introduced into apricot. The plum pox virus leads to the feared Sharka disease, for which there is no cure. The resulting transgenic tree shows a markedly decreased sensitivity to this virus. The reason why continuous exposure of the tree to the viral coat protein leads to tolerance against viral infection is not yet understood, however. Thus, now there are a number of different techniques to introduce foreign genes into plants.

Essentially all major crop plants can be (and have been or are being) genetically engineered, the procedures are now routine and the frequency of success is very high. Even though genetically engineered crops are more costly than the usual ones, they have been rather readily accepted by US farmers provided that tangible benefits can be demonstrated. However, it is questionable whether the farmer in poorer countries can come up with the funds to "try out" and use the new crops.

Another issue in this respect is how genetically engineered crops are perceived by the consumer. Even though in the US there is little resistance to such crops as long as the products can be shown to be safe and advantageous, in other countries genetically modified foods are received poorly by the consumer. It is unlikely that there is a rationally sound basis for this rather hostile reaction of the consumer, as most of the crops are the result of human manipulation and may have been treated with harmful herbicides and pesticides.

Time and education will need to be invested to provide consumers and consumer advocates with a balanced opinion on the acceptability of the origin of their foods. One area of particular concern for some people is the lack of labeling of genetically engineered foods, and legislation may be introduced to address this issue. On the other hand, as so many plants (soybean, corn, etc.) are genetically modified and the nature of the genetic modification is not necessarily easy to explain, it may be simpler to label those foods that are guaranteed free of "genetically modified organisms" or their products. However, keep in mind that essentially all agricultural products have been genetically modified by traditional breeding, so it may be difficult to define what is actually free of genetically modified organisms.

## BIOFORTIFICATION OF FOOD CROPS

Biofortification of food crops makes sense as part of an integrated food-systems approach to reducing malnutrition. It addresses the root causes of micronutrient deficiencies, targets the poorest people, and is scientifically feasible and cost-effective. It is a first step in enabling rural households to improve family nutrition and health in a sustainable way. HarvestPlus is a coalition of CGIAR Future Harvest Centres or Institutes, partner collaborating institutions and supportive donors. The International Centre for Tropical Agriculture and IFPRI are coordinating the plant breeding, human nutrition, crop dissemination, policy analysis and impact activities to be carried out at international Future Harvest Centres, national agricultural research and extension institutions, and departments of plant science and human nutrition at universities in both developing and developed countries.

An initiative of the Consultative Group on International Agricultural Research, HarvestPlus is a global alliance of research institutions and implementing agencies coming together to breed and disseminate crops with improved nutritive value, *e.g.* with a higher content of iron, zinc and vitamin A. The biofortification approach is backed by sound science.

*Research on this funded by the Danish International Development Assistance and coordinated by the International Food Policy Research Institute led to the following conclusions*:

- Substantial, useful genetic variation exists in key staple crops;
- Breeding programmes can readily manage nutritional quality traits, which for some crops have proven to be highly suitable and simple to screen for;
- Desired traits are sufficiently stable across a wide range of growing environments; and
- Traits for high nutrition content can be combined with superior agronomic traits and high yields.

Initial biofortification efforts will focus on six staple crops for which prebreeding studies have been completed: beans, cassava, sweet potatoes, rice, maize and wheat. The potential for nutrient enhancement will also be studied in ten additional crops that are important components in the diets of those with micronutrient deficiencies: bananas/plantains, barley, cowpeas, groundnuts, lentils, millet, pigeon-peas, potatoes, sorghum and yams. During the first four years of the project, the objectives are to: determine nutritionally optimal breeding objectives; screen CGIAR germplasm for high iron, zinc and beta-carotene amounts; initiate crosses of high-yielding adapted germplasm for selected crops; document cultural and food-processing practices, and determine their impact on micronutrient content and bioavailability; identify the genetic markers available to facilitate the transfer of traits through conventional and novel breeding strategies; carry out in-vitro and animal studies to determine the bioavailability of the enhanced micronutrients in promising lines; and initiate bio-efficacy studies to determine the effect on biofortified crops on the micronutrient status of humans.

During the following three years, the objectives are to: continue bio-efficacy studies; initiate farmer-participatory breeding; adapt high-yielding, conventionally-bred, micronutrient-dense lines to select regions; release new conventionally-bred biofortified varieties to farmers; identify gene systems with potential for increasing nutritional value beyond conventional breeding methods; produce transgenic lines at experimental level and screen for micronutrients, test for compliance with biosafety regulations; develop and implement a marketing strategy to promote the improved varieties; and begin production and distribution.

During the last three years of the project, production and distribution of the improved varieties will be scaled up; the nutritional effectiveness of the programme will be determined; and the factors affecting the adoption of biofortified crops, the health effects on individuals and the impact on household resources will be identified.

### *Rice*

Rice is the dominant cereal crop in many developing countries and is the staple food for more than half of the world's population. In several Asian countries, rice provides 50 per cent to 80 per cent of the calorie intake of the poor. In South and South-East Asian countries, more than half of all women and children are anaemic; increasing rice nutritive value can therefore have significant positive health impact. Food-consumption studies suggested that doubling the iron content in rice could increase the iron intake of the poor by 50 per cent; germplasm screening indicated that a doubling of iron and zinc content in unmilled rice was feasible. Milling losses vary widely by rice variety, with losses of iron being higher than losses of zinc, which suggests than more zinc is deposited in the inner parts of the rice endosperm. Under the HarvestPlus project, improved rice germplasm will be provided to national partners in Bangladesh, Indonesia, Vietnam, India and the Philippines.

The improved features will be incorporated into well-adapted and agronomically-preferred germplasm in ongoing breeding programmes at the national and regional level. A plant-biotechnology approach is the current priority for enhancing provitamin-A content of the rice endosperm. The leading varieties will be field tested for agronomic performance and compositional stability in at least four countries.

### *Wheat*

The International Maize and Wheat Improvement Centre is leading the HarvestPlus research endeavour on wheat biofortification in order to increase people's intake of iron and zinc. Given that spring wheat varieties developed by CIMMYT and its partners are used in 80 per cent of the global spring wheat area, the potential impact of iron-enhanced wheat could be dramatic. The initial target countries will be Pakistan and India, in the area around the Indo-Gangetic plains, a region with high population densities and high micronutrient malnutrition.

The highest contents of iron and zinc in wheat grains are found in landraces of wild relatives of wheat such as Triticum dicoccon and Aegilops tauschii. Because these wild relatives of wheat cannot be crossed directly with modern wheat, researchers facilitated the cross between a high-micronutrient wild relative, Aegilops tauschii, and a high-micronutrient primitive wheat, Triticum dicoccon, to develop a variety of hexaploid wheat that can be crossed directly with current modern varieties of wheat and have 40 per cent to 50 per cent higher contents of iron and zinc in the grain than modern wheat.

The first biofortified lines will be delivered to the target region by 2005, *i.e.* broadly-adapted, high-yielding, disease-resistant wheat lines. The first high-yielding lines with confirmed iron and zinc contents in the grain should be available for regional deployment by mid-2007. Researchers will be exploring the introduction of the ferritin gene in wheat and will establish the feasibility of increasing the concentration of iron and zinc in the grain using advanced biotechnology approaches in addition to conventional plant breeding. Molecular markers for the iron and zinc genes that control concentration in the grain were being identified in order to facilitate their transfer. Scientists will also carry out studies on bioavailability to determine the extent to which iron and zinc status in animal and human subjects is improved when biofortified varieties are consumed on a daily basis over several months.

### *Maize*

Maize is the preferred staple food of more than 1.2 billion consumers in sub-Saharan Africa and Latin America. Over 50 million people in these regions were vitamin A-deficient in 2004. The International Maize and Wheat Improvement Centre and the International Institute of Tropical Agriculture are identifying micronutrient-rich maize varieties and will carry out adaptive breeding for local conditions in partnership with National Agricultural Research Systems in Africa and Latin America.

The project under HarvestPlus is initially focusing on maize varieties having increased contents of provitamin A because a useful range of genetic variation has already been identified for this trait. The first target countries are Brazil, Guatemala, Ethiopia, Ghana and Zambia. To support the breeding programme, research is being conducted in Brazil, the USA and Europe to develop simple, inexpensive and rapid screening protocols for provitamin A, so as to reduce the cost of assays from $70-100 to $5-10 per sample. Research in Brazil and the USA is also focused on finding genetic markers to facilitate marker-assisted selection for provitamin A concentration.

In collaboration with the University of Wageningen, a human efficacy trial was planned with provitamin A-rich maize in Nigeria for 2005 in order to study provitamin A retention or loss for different storage, processing and common cooking methods. To facilitate extension and dissemination of biofortified maize varieties, country teams will be formed in the target countries in order to conduct adaptive breeding research, farmer-participatory variety evaluations, nutritional advocacy and promotional activities.

### *Beans*

Common beans are the world's most important food legume, far more so than chickpeas, faba beans, lentils and cowpeas. For more than 300 million people, an inexpensive bowl of beans is the main meal of their daily diet. The focus of HarvestPlus research is on increasing the concentration of iron and zinc in agronomically superior varieties. Over 2,000 accessions from the International Centre for Tropical Agriculture gene-bank and several hundred collections of African landraces have been screened for their nutrient contents. While the average iron concentration in these varieties is about 55 mg per kg, researchers have found varieties the content of which exceeds 100 mg per kg. The eventual goals are to obtain favourable combinations for productivity and nutritional traits, double the iron concentration and increase zinc concentration by about 40 per cent. The first bred lines with 70 per cent higher iron will likely emerge in 2006, while lines with double concentration of iron are anticipated in 2008.

The proportion of iron and zinc that can be absorbed from legumes such as common beans is typically low, due to anti-nutrients, specifically phytates and polyphenols, which normally bind to the iron and zinc, making them unavailable to the organism. Research indicated that it might be possible to reduce polyphenol concentrations genetically, thereby improving iron bioavailability. In contrast, vitamin C is an iron-absorption enhancer because it binds to iron and prevents it from becoming attached to the iron-absorption inhibitors. Beans are often consumed with vegetables, including bean leaves with the potential of bean leaves as a source of vitamin C still to be explored.

### *Cassava*

Cassava, also known as manioc or tapioca, is a perennial crop native of tropical America that is also widely consumed in sub-Saharan Africa and parts of Asia. With its productivity on marginal soils, ability to withstand disease, drought and pests, flexible harvest dates, cassava is a remarkably adapted crop consumed by people in areas where drought, poverty and malnutrition are often prevalent. Cassava is typically white in colour and, depending on the amounts of cyanogenic compounds, can be sweet or bitter. The International Centre for Tropical Agriculture will coordinate HarvestPlus' overall activities on cassava biofortification and be primarily responsible for research in Asia, Latin America and the Caribbean.

The IITA will be responsible for cassava biofortification in Africa. In collaboration with the University of Campinas, São Paulo State, Brazil, the total content of provitamin A in cassava varieties will be determined spectrophotometrically. Provitamin-A retention studies will also be carried out on different preparation and cooking methods used in cassava-consuming countries. A method for storing cassava roots for several weeks or a few months is needed for programmes quantifying hundreds of samples per year. Initial data suggest that the anti-oxidant property of a few yellow pigments in cassava roots may delay physical deterioration of the roots.

The longer shelf life of yellow cassava roots may not only appeal to farmers and consumers, but may also increase the demand for biofortified varieties. Nutritionally improved germplasm coupled with superior agronomic performance can be developed as a medium-term approach with products reaching the farmers as soon as 2009. The aim is to identify and select, from the varieties having both high provitamin-A contents and good agronomic performance, those with the highest iron and/or zinc content.

### *Sweet Potato*

Sweet potato is an important part of the diet in East and Central Africa where vitamin-A deficiency is widespread. At present, African predominant sweet potato cultivars are white or yellow-fleshed varieties that contain small amounts of provitamin A. In contrast, the orange-fleshed varieties are believed to be one of the least expensive, rich, year-round sources of provitamin A. Boiled orange-fleshed sweet potato, such as the Resisto variety developed in South Africa, contains between 1,170 and 1,620 Retinol Activity Equivalents per 100 g and is estimated to provide between 25 per cent and 35 per cent of the recommended daily allowance for a preschool child. Experts at the International Potato Centre who developed a biofortified orange-fleshed sweet potato, estimated that when fully disseminated, this sweet potato could reduce vitamin-A deficiency in as many as 50 million children. To encourage a switch from non-orange to orange-fleshed varieties, the texture of the latter must be changed because they tend to have a high-moisture content and adults prefer varieties with a low water content, *i.e.* a high dry biomass. Plant breeding is ongoing to increase the dry biomass of the provitamin A-rich orange varieties, to improve organoleptic characteristics and at the same time improve their resistance to viruses and drought. About 40 varieties of sweet potato with high dry biomass and provitamin-A content have been introduced to sub-Saharan Africa. Of these, 10 to 15 were being tested widely in different agro-ecological areas in some countries.

Some original varieties, mainly local landraces, have been well accepted by farmers and were being distributed on a small scale. HarvestPlus' biofortification activities in sweet potato will be initially focused on Ethiopia, Ghana, Kenya, Mozambique, Rwanda, South Africa, Tanzania and Uganda. The variation in provitamin-A content of newly harvested roots can be as much as 45 per cent. Much of the provitamin A appears to be retained during storage, food preparation and cooking. In the South African Resisto variety, the provitamin-A activity of the boiled roots was between 70 per cent and 80 per cent of that of freshly harvested roots. Additional studies were to be carried out in 2004 to determine the provitamin-A losses during food processing and cooking based on the usual practices found in East and Central Africa. A human bioefficacy study using an organoleptically acceptable promising variety was planned for 2005, once the food processing studies were completed.

The $100-million ten-year HarvestPlus programme will be financed during the first four years mainly by the World Bank, the USAID and DANIDA. The Bill and Melinda Gates Foundation would contribute $25 million towards the total cost of the programme. In addition, the Canadian Agency for International Development will allocate funds for the Latin American part of the programme.

## GENETICALLY MODIFIED PLANTS AND CROPS

Genetically modified plants are created by the process of genetic engineering, which allows scientists to move genetic material between organisms with the aim of changing their characteristics. All organisms are composed of cells that contain the DNA molecule. Molecules of DNA form units of genetic information, known as genes. Each organism has a genetic blueprint made up of DNA that determines the regulatory functions of its cells and thus the characteristics that make it unique. Prior to genetic engineering, the exchange of DNA material was possible only between individual organisms of the same species. With the advent of genetic engineering in 1972, scientists have been able to identify specific genes associated with desirable traits in one organism and transfer those genes across species boundaries into another organism.

A gene from bacteria, virus, or animal may be transferred into plants to produce genetically modified plants having changed characteristics. Thus, this method allows mixing of the genetic material among species that cannot otherwise breed naturally. The success of a genetically improved plant depends on the ability to grow single modified cells into whole plants. Some plants like potato and tomato grow easily from single cell or plant tissue. Others such as corn, soy bean, and wheat are more difficult to grow. After years of research, plant specialists have been able to apply their knowledge of genetics to improve various crops such as corn, potato, and cotton. They have to be careful to ensure that the basic characteristics of these new plants are the same as the traditional ones, except for the addition of the improved traits. The world of biotechnology has always moved fast, and now it is moving even faster. More traits are emerging; more land than ever before is being planted with genetically modified varieties of an ever-expanding number of crops. Research efforts are being made to genetically modify most plants with a high economic value such as cereals, fruits, vegetables, and floriculture and horticulture species.

## Biological Cell

It only takes one biological cell to create an organism. In fact, there are countless species of single celled organisms, and indeed multi-cellular organisms like ourselves.

A single cell is able to keep itself functional by owning a series of '*miniature machines*' known as *organelles*. The following list looks at some of these organelles and other characteristics typical of a fully functioning cell.

- *Mitochondrion:* An important cell organelle involved in respiration
- *Cytoplasm:* A fluid surrounding the contents of a cell and forms a vacuole
- *Golgi Apparatus:* The processing area for the creation of a glycoprotein
- *Endoplasmic Reticulum:* An important organelle heavily involved in protein synthesis.

- *Vesicles: Packages* of substances that are to be used in the cell or secreted by it.
- *Nucleus*: The "brain" of a cell containing genetic information that determines every natural process within an organism.
- *Cell Membrane:* Also known as a plasma membrane, this outer layer of a cell assists in the movement of molecules in and out the cell plays both a structural and protective role
- *Lysosomes:* Membranous sacs that contain digestive enzymes

## Cell Wall

A structure that characteristically is found in plants and prokaryotes and not animals that plays a structural and protective role.

## Cell Specialisation

Cells can become specialised to perform a particular function within an organism, usually as part of a larger tissue consisting of many of the same cells working in tandem:

- *Nerve cells* to operate as part of the nervous system to send messages back and forth via the brain at the centre of the nerve system.
- Skin cells for waterproof protection and protection against pathogens in the open air environment.
- *Xylem* tubes to transport water around plants and to provide structural support for the plant as a whole.

Cells combine their efforts in these tissue types to perform a common cause. The task of the specialised cell will determine in what way it is going to be specialised, because different cells are suited to different purposes, as illustrated in the above list:

- Muscle cells are long and smooth in structure and their elastic nature allows these cells to perform flexible movements, just as they do in our own body's.
- Some *white blood cells* contain powerful digestive enzymes to eliminate pathogens by breaking them down to the molecular level.
- Cells at the back of the eye are sensitive to light stimuli, and thus can *interpret* differences in light intensity which can in turn be interpreted by our nervous system and brain.

Many of these cells contain organelles, though after some cells are specialised, they do not possess particular characteristics as they do not require them to be there. *i.e.* efficiency is the key, no resources are wasted and the resources available are put to their idyllic optimum.

### The Cell Membrane

The *cell membrane,* otherwise known as the plasma membrane is a semi-permeable structure consisting mainly of *phospholipid* (fat) molecules and proteins. They are structured in a *fluid mosaic model,* where a double layer of phospholipid molecules provide a barrier accompanied by proteins.

It is present round the circumference of a cell to acts as a barrier, keeping foreign entities out the cell and its contents (like cytoplasm) firmly inside the cell.

The plasma membrane allows only selected materials to pass in and out of a cell, and is thus known as a selectively permeable membrane.

### Cell Transport

There are three methods in which ions are transported through the cell membrane into the cell,

- *Active Transport*: Active transport is the transport of molecules with the active assistance of a carrier that can transport the material against a natural *concentration gradient.*
- *Passive Transport (Diffusion):* The movement of molecules from areas of high concentration (*i.e.* outside a cell) to areas of low concentration (*i.e.* within a cell) via a carrier. This process does not require energy.
- *Simple Diffusion:* The movement of molecules from areas of high concentration to areas of low concentration in a free state. *Osmosis* of water involves this type of diffusion through a selectively permeable membrane (*i.e.* plasma membrane)

## GENETIC TAGGING OF AQUACULTURAL SPECIES

Species-specific DNA markers can be used to identify animal species such as commercial molluscs and crustaceans, which represent a high proportion of aquacultural species. For instance, in Thailand, at the National Centre for Genetic Engineering and Biotechnology Marine Biotechnology Unit, species-specific markers based on 16S ribosomal DNA polymorphism have been developed for penaeid shrimps, tropical abalone and oysters. The black tiger shrimp is the most commercially important cultured species in Thailand. Because of outbreaks of diseases, the white shrimp has been introduced into Thailand and cultured commercially.

On the other hand, external characteristics of P. monodon and P. semisulcatus are similar, but the growth rate of the latter is approximately three times slower than that of the former. In addition, P. merguiensis larvae, which could not yet be successfully cultured, were sold as those of P.vannamei. Species-specific markers were therefore developed for identifying the aforementioned species and P. japonicus as well. These markers can be applied to ensure quality control by properly labeling traded shrimp larvae. Three species of tropical abalone are found in Thailand's waters: Haliotis asinina, H. ovina and H. varia. However, H. asinina is the most productive one, as it provides the highest ratio between meat weight and total weight. H. asinina specific markers based on 16S rDNA polymorphism have been developed in order to prevent supplying the wrong abalone larvae for the industry as well as to foster quality control of abalone products from Thailand.

Oyster farming has shown rapid growth in Thailand over the last few years. Taxonomic difficulties relating to Thai oysters have had a limiting effect on the culture efficiency and development of their closed life cycle. Molecular genetic markers have therefore been developed to identify the three commercially cultured oysters, Crassostrea belcheri, C. iredalei and Saccostrea cucullata.

## DNA Fingerprinting of Grapevine Varieties

Since 1990, US and French researchers have been trying to establish the phylogenetic tree of the varieties of grapevine grown throughout the world, using the fingerprinting technique. The research consists of analyzing the structure of certain regions of the genome of some grapevine varieties and comparing it with that of other varieties, so as to establish possible phylogenetic relations. For this analysis, the DNA is extracted from young ground leaves, but also from fruits and branches. The results of this first research have been published in 1999 in the Science journal.

There are about 6,000 grapevine varieties cultivated worldwide. In order to establish the origin of a variety, specialists used to rely on phenotypic traits, such as the morphology of leaves, berries and grapes. In this way, varieties could be grouped in a few families. DNA fingerprinting enables the researchers to go further. It appears that the current grapevine varieties are the remote offspring of the grapevines grown during the Antiquity around the Mediterranean, or during the European Middle Ages. The current varieties are the result of lengthy breeding activities, identification and comparison work, and stabilization of the selected strains or lines. The team led by C. Meredith and J. Bowers of the Department of Viticulture and Oenology of the University of California, Davis, has confirmed that the cabernet sauvignon variety–which is dominant in the Medoc region of France and is at the origin of most red grapevine varieties in the New World–was in fact the offspring of the cabernet franc and sauvignon, two varieties deeply entrenched in the middle valley of the Loire river.

The cooperation between the Californian team and the French specialists of the National Higher School of Agronomy, Montpellier, associated with the Genetic Research and Breeding Unit of the National Agricultural Research Institute has led to undisputable results concerning the origins of various grapevine varieties: the chardonnay, aligote, gamay and melon of Burgundy are all cousins and almost siblings. Such a conclusion, derived from the analysis of DNA fingerprinting, was not a surprise for oenologists and tasters because the wines made from these four varieties share common structures and aromas. This kinship is particularly expressed in the ageing wines. Similarly, aged wines of the chenin variety resemble the Hungarian tokay. But more surprising than the discovery of that kinship among the four grapevine varieties, was the identity of one of the progenitors of the initial couple that gave rise to these varieties. Indeed, several historical elements were in favour of the creation of lines through the cross pollination between the pinot noir and the white gouais; these crosses have given birth–as proven by the fingerprinting analysis–to the three white and red varieties grown for a long time in various French provinces, and for some decades, in many regions of the globe.

The surprising aspect of this discovery was that the white gouais is almost unknown, although the vine specialists in Montpellier continue to grow it and to make wine form it for their own pleasure. However, this variety has played a key role in the origin of French viticulture, just as to R. Dion in his Histoire de la vigne et du vin en France des origines au XIXe siècle. A document dated from the 12th century mentioned this variety as a lower-grade one; in 1338, the white gouais was found in Metz under the name of goez; at that time, instructions were given to eliminate this variety from all the Metz territory and to privilege only the white and black fromental, considered as higher-grade varieties.

The gouais was found in Paris during the 14th century and, owing to the expansion of the workers' population, it progressively replaced the pinot noir of Burgundy, which was a good variety of Parisian vineyards. The extension of the gouais was due to the wish of the winemakers to produce a cheaper wine. However, the phenomenon was limited to Paris and its suburbs; in the vineyards located away from the capital, the gouais was rejected, more noble grapevine varieties were used and contributed to the reputation of French viticulture.

The white gouais was also formerly grown in the Jura and Franche-Comté. For the US and French researchers, this variety which has played a key role in the history of vine and wine, is the same as the heunisch variety of Central Europe, introduced in Gaul by a Roman emperor originating from Dalmatia. In Montpellier, the French researchers participating in the joint study with the US scientists from the University of California, Davis, have tried to reproduce the breeding between the pinot noir and white gouais in order to seek confirmation of the genetic research. Other attempts were expected to widen even more the range of cultivated grapevine varieties, for both their fruits and wines derived from them. But this approach was hindered by a drastic regulation, which practically prohibits any venture of this kind, while non-French winemakers and vinegrowers could do it.

In Apulia, in the heel of the Italian boot, drawing on grapes grown by up to 1,600 small farmers in the area, a California wine consultant associated with another Italian wine consultant from Friuli are producing and marketing wines that have scored a great success worldwide, with 2004 projected sales 15 times as big as those in 1998, the winery's first year. The wines are called A-Mano–handmade–and by far the best known is a robust red made from a once-obscure grape named primitivo. DNA testing by Carole Meredith at the University of California, Davis, established that primitivo is a descendant of a grape called crljenak kastelanski, widely known in the 18th and 19th centuries on the Dalmatian coast of Croatia.

California's zinfandel, she showed, is genetically the same as primitivo, though how it crossed the ocean remains a subject of dispute. Apulian primitivo and zin are not twins, of course; climate, soil and vinification all help to shape a wine's look, aroma and flavour, along with the grape variety. But the two share several characteristics: both are fruit-rich, chewy, sometimes lush wines, a deep violet-red in colour, often too high in alcoholic content for comfort, but much more subtle if carefully handled.

For years, primitivo was used to add unacknowledged heft to chianti, barbaresco and even red burgundy. Nowadays, primitivo can stand on its own feet. In addition to A-Mano primitivo, other high-quality primitivos are grouped in an organization called the Academia dei Racemi, not a true cooperative but an association in which each member makes his own wine and joins the others for marketing support and technical advice. Based in Manduria, between the old cities of Taranto and Lecce, the group includes value-for-money labels like Masseria Pepe, Pervini and Felline.

# 3

# Plant Tissue Culture: Teaching and Techniques

The process of explant introduction into in vitro conditions depends mainly on the disinfecting phase. Problems in handling produce contamination of the in vitro plant and so, a similar disinfection process to that of the in vitro introduction technique is recommended.

However, it is possible to use antibiotics included in the medium, but just temporarily, while the plantlets are growing. Among the bacterial contaminants are Bacillus sp., Erwin/a sp., Pseudomonas sp., etc. In the case of systemic infections, the use of disinfectants is not effective, because the pathogens are located in the vascular system. In this case, it is recommended to use specific antibiotics and meristems or bud cuttings.

On the other hand, the process of in vitro introduction involves the use of explants in different physiological stages, as dormant buds, which require growth regulators for example, Gibberellic acid) to stimulate and accelerate bud growth.

During the process of in vitro introduction there is a close connection between the greenhouse and the laboratory; that is why the necessary precautions should be taken to avoid the entrance of contaminants into the laboratory.

- In the greenhouse
  - Take cuttings from the mother plant, or buds from greenhouse tuber sprouts.
- In the laboratory
  - Immerse the cuttings for 10 minutes in a beaker conveniently labeled with the accession name, and containing a solution of 0.5 per cent acaricide with three drops/I of Tween 20.

- Throw away the acaricide solution and rinse the nodes with tap water
- Prepare the laminar flow chamber.
- Immerse the cuttings in 700/o alcohol for 30 seconds and take the beakers immediately to the laminar flow chamber

• In the laminar flow chamber
  - Eliminate the alcohol and replace it with a 2.50/o solution of calcium hypochlorite. Keep the cuttings immersed during 15 minutes.
  - Throw away the hypochlorite, and rinse the cuttings three times with sterile water
  - Keep the nodes immersed in sterile water until bud extraction.
  - Dissect the buds on a sterile plate and place them in the culture medium.

## BASIC STEPS OF PLANT TISSUE CULTURE

Some basic steps are followed in plant tissue culture which is described below.

### Explant Selection

Explant is the plant fraction used for tissue culture. Explant from healthy and young part of the plant is used. Parenchymas from stems, rhizomes, tubers, roots are easily accessible and respond quickly to culture condition.

### Sterilisation Elimination

Sterilisation means elimination of all living organisms (microbes). The tissue culture is carried out in completely aseptic condition for which there is a need to properly sterilise all the glasswares culture media and explants.

The explants are surface sterilised by repeated washing in sterile water and by using disinfectants such as mercuric chloride, hydrogen peroxide, etc. The glassware, culture media and other instruments are sterilised in an instrument called autoclave where sterilisation is done in steam under high pressure or in a pressure cooker.

### Preparation of Nutrient or Culture Medium

The culture medium is prepared in aseptic condition. The basic constituents of any culture medium are:

- *Inorganic nutrients:* Inorganic nutrients include macronutrients as salts of nitrogen, phosphorous, potassium, calcium, magnesium and sulphur and micronutrients like boron, molybdenum, copper, zinc. Iron and chloride

## Source of Carbon

Sucrose is used as the source of carbon. Several growth hormones like 2,4,1), cytokinins-benzylaminopurine, kinetin, myoinositol, 1AA, XAA arc used, vitamins like nicosstinic acid, pyridoxinc-IICL are added to the medium. After the addition of all the ingredients in appropriate proportion, agar-agar is added to prepare a solid medium. In some types of culture (like root culture) liquid medium is used (no agar-agar used).

## Process of Inoculation

Inoculation is the process of the transfer of explant to suitable nutrient medium contained in culture vessels. This is done in sterile condition either in an inoculation chamber or laminar flow. After the inoculation the culture vessels are maintained in controlled temperature and light. The suitable temperature, for tissue culture ranges between 18-25°C.

## Callus Configuration and its Culture

A callus is an amorphous mass of loosely arranged thin walled parenchyma cells developing from proliferating cells of parent tissue (Dodos and Roberts, 1985). The nutrient medium supplemented with auxins induces cell division and soon the upper surface of explant is covered by callus.

The callus has the biological potential to develop normal root, shoots and ultimately forming a plant. Callus is formed through three developmental stages: induction, cell division and differentiation. Callus formation is governed by the source of explant, nutritional composition of the medium and environmental factors.

During the induction period the metabolic rate of cells is stimulated. Owing to increased metabolic rate cells enter to cell division stage. In the third phase, cellular differentiation and expression of certain metabolic pathways start leading to secondary products. When callus are grown on a nutrient medium for a long time, it becomes essential to subculture it in fresh media.

## Development of Organogenesis

Organogenesis means the development of organs like root, shoot and leaves (but not embryo). Organogenesis starts with stimulation caused by the chemical of medium, substances carried over from the original explants and endogenous compounds produced by the culture. Organogenesis can be induced with the application of varying proportions of auxin and cytokinin. Skoog and Miller (1957) demonstrated that a high ratio of auxin: cytokinin (3: 0.02) stimulated root formation in tobacco callus but a low ratio of the same (3: 0,2) induced shoot formation.

## Formation of Somatic Embryogenesis

This is the process of inducing embryo formation from somatic cells of cultured plant tissue. The embryo thus developed is knowTi as embryoids. Two different nutritional media are required to obtain embryoids. First medium contains auxin to initiate embryogenic cells. Second medium lacks auxin or has reduced level of auxins for subsequent development of embryonic cells into embryoids and plantlets.

The embryogenic cells pass through three different stages, *e.g.*, globular, heart shaped and torpedo shaped to form embryoids. Some plants in which somatic embryogenesis has been induced *in vitro* are Atropa belladona, Brassica oleracea, Carica papaya, Coffea arabica, Citrus cinensis, Daucus carroia, Nicotiana tobacum, etc.

Somaclonal Variations: In 1981, P.J Larkins and W. R. Scowcrott at the Division of Plant Industry, C.S.I.R.O. Australia gave the name somaclonal variation to genetic variability generated during tissue culture.

Though the cultured tissues are grown from single explant, over a long period of maintenance genetic variabilities are marked in the cultures.

*This may be due to:*

- Reflection of heterogeneity between cells and explant tissues,
- A simple representation of spontaneous mutation rate or
- Activation by culture environment of transposition of genetic materials.

Somaclonal variants of leaf callus culture of Solanum tuberosum have shown characters like disease resistance, variations in maturity dates of tubers, yield and shape.

Cell suspension cultures: Cell suspension is prepared by transferring a fragment of callus (about 500 mg) to liquid medium (500 ml) and agitating them aseptically in a shaker to make the cells free. The suspension then includes single cell, cell aggregates, and residual including dead cells. A good suspension contains high proportion of single cells.

*Cell suspension cultures have many advantages over the callus cultures as below:*

- The cell suspension can be pipetted.
- They are less heterogeneous.

- They can be cultured in volumes up to 1,500 litres.
- They can be subjected to more stringent environmental controls.
- They can be manipulated for production of natural products.

## CULTURED CELLS AND TISSUES

The large scale commercial propagation of plant material based on plant tissue culture was pioneered in the USA. During the last thirty years, tissue culture-based plant propagation has emerged as one of the leading global agro-technologies. Between 1986 and 1993, the worldwide production of tissue cultured plants increased 50 per cent.

In 1993, the production was 663 million plants. By 1997, production had risen to 800 million plants. During 1990–1994, the micropropagation industry declined in Europe, mainly due to production shifting to developing countries, but since then because of the demand for high quality and number, production in European countries has increased.

Since 1995, production has increased by 14 per cent in Asian countries, mainly due to the market entry of China, while the increase in South and Central America was from production in Cuba. More recently, some companies from Israel, the USA and UK have shifted their production requirements to Costa Rica and India.

Tissue cultured plants have as yet to reach many growers and farmers in the developing countries. The primary advantage of micropropagation is the rapid production of high quality, disease-free and uniform planting material.

The plants can be multiplied under a controlled environment, anywhere, irrespective of the season and weather, on a year-round basis. Production of high quality and healthy planting material of ornamentals, and forest and fruit trees, propagated from vegetative parts, has created new opportunities in global trading for producers, farmers, and nursery owners, and for rural employment.

Micropropagation technology is more expensive than the conventional methods of plant propagation, and requires several types of skills. It is a capital-intensive industry, and in some cases the unit cost per plant becomes unaffordable. The major reasons are cost of production and know-how. During the early years of the technology, there were difficulties in selling tissue culture products because the conventional planting material was much cheaper.

Now this problem has been addressed by inventing reliable and cost effective tissue culture methods without compromising on quality. This requires a constant monitoring of the input costs of chemicals, media, energy, labour and capital. In the industrialized countries, labour is the main factor that contributes to the high cost of production of tissue-cultured plants.

To reduce such costs, some steps can be partially mechanized, *e.g.* use of a peristaltic pump for medium dispensing, and of dishwashers for cleaning containers. In the less developed countries of Africa, Asia, and Latin America, where labour is relatively cheaper, consumables such as media, culture containers, and electricity make a comparatively greater contribution to production costs.

For example, the cost of medium preparation (chemicals, energy and labour) can account for 30–35 per cent of the micropropagated plant production. However, automated production processes based on pre-sterilized membrane capsules, bioreactors, mechanized explant transfer, and container sealing are not commercially viable propositions in many developing countries. Therefore, low cost alternatives are needed to reduce production cost of tissue-cultured plants.

Many of the low cost technology options described in this publication can be incorporated in various steps of plant micropropagation. The occasional tissue culturegenerated variants (somaclones) and rare spontaneous bud mutants as well as those obtained from induced mutations can also be propagated by deployment of low cost techniques described.

Micropropagation, in combination with radiation-induced mutations, speeds up the recovery, multiplication and release of improved varieties in vegetatively propagated plants. Hence, low cost technology will be of great value for large scale plant multiplication of mutants of many fruits, shrubs, flowers and forest trees that are conventionally vegetatively propagated.

## Plant Tissue Culture

Plant tissue culture refers to growing and multiplication of cells, tissues and organs of plants on defined solid or liquid media under aseptic and controlled environment. The commercial technology is primarily based on micropro-pagation, in which rapid proliferation is achieved from tiny stem cuttings, axillary buds, and to a limited extent from somatic embryos, cell clumps in suspension cultures and bioreactors. The cultured cells and tissue can take several pathways. The pathways that lead to the production of true-to-type plants in large numbers are the preferred ones for commercial multiplication.

The process of micropropagation is usually divided into several stages *i.e.*, prepropagation, initiation of explants, subculture of explants for proliferation, shooting and rooting, and hardening. These stages are universally applicable in large-scale multiplication of plants. The delivery of hardened small micropropagated plants to growers and market also requires extra care.

Plant tissue culture refers to growing and multiplication of cells, tissues and organs on defined solid or liquid media under aseptic and controlled environment. Plant tissue culture technology is being widely used for large-scale plant multiplication. The commercial technology is primarily based on micropropagation, in which rapid proliferation is achieved from tiny stem cuttings, axillary buds, and to a limited extent from somatic embryos, cell clumps in suspension cultures and bioreactors.

## Explant Source

Plant tissue cultures are initiated from tiny pieces, called explants, taken from any part of a plant. Practically all parts of a plant have been used successfully as a source of explants. In practice, the "explant" is removed surgically, surface sterilized and placed on a nutrient medium to initiate the mother culture, that is multiplied repeatedly by subculture. The following plant parts are extensively used in commercial micropropagation. Shoot-tip and meristem-tip culture: Shoots develop from a small group of cells known as shoot apical meristem.

The apical meristem maintains itself, gives rise to new tissues and organs, and communicates signals to the rest of the plant. Shoot-tips and meristem-tips are perhaps the most popular source of explants to initiate tissue cultures. The shoot apex explant measures between 100 to 500 μ m and includes the apical meristem with 1 to 3 leaf primordia. The apical meristem of a shoot is the portion lying distal to the youngest leaf primordium, and is ca.100 μ??m in diameter and 250μ?m in length with 800-1200 cells. In practice, shoot-tip explants between 100 to 1000 μ m are cultured to free plants from viruses. Even explants larger than 1000 μ m have been frequently used. The term "meristem-tip culture" has been suggested to distinguish the large explants from those used in conventional propagation.

Nodal or axillary bud culture: This consists of a piece of stem with axillary bud culture with or without a portion of shoot. When only the axillary bud is taken, it is designated as "axillary bud" culture. Floral meristem and bud culture: Such explants are not commonly used in commercial propagation, but floral meristems and buds can generate complete plants. Other sources of explants: In some plants, leaf discs, intercalary meristems from nodes, small pieces of stems, immature zygotic embryos and nucellus have also been used as explants to initiate cultures.

Cell suspension and callus cultures: Plant parts such as leaf discs, intercalary meristems,-stem-pieces, immature embryos, anthers, pollen, microspores and ovules have been cultured to initiate callus. A callus is a mass of unorganized cells, which in many cases, upon transfer to suitable medium, is capable of giving rise to shoot-buds and somatic embryos, which then form complete plants.

Such calli on culture in liquid media on shakers are used for initiating cell suspensions. Liquid suspension cultures maintained on mechanical shakers achieve fast and excellent multiplication rates. However, in commercial micropropagation, calli are cultured mostly in bottles and flasks kept on semi-solid or liquid media. To a limited extent, bioreactors have become popular for somatic embryogenic cultures. It is considered that some day robotics could be adapted to bioreactorbased micropropagation.

## PLANT TISSUE CULTURE MEDIA

Plant Tissue Culture refers to the technique of growing plant cells, tissues, organs, seeds or other plant parts in a sterile environment on a nutrient medium.

*Culture media used for in vitro cultivation of plant cells are composed of following basic components*:

- *Complex mixture of salts*: Essential elements, or mineral ions.
- *Organic supplement*: Vitamins and/or amino acids.
- *Carbon source*: Usually sugar sucrose.
- Gelling agents
- Plant Growth Regulators
- Antibiotics

### Complex Mixture of Salts

These include essential elements or mineral ions important for plant nutrition and their physiological function.

*The essential elements can further be divided into the following categories*:

- Macroelements (or macronutrients)
- Microelements (or micronutrients)
- Iron source

### Macroelements

Elements are required in large amounts for plant growth and development. Nitrogen, phosphorus, potassium, magnesium, calcium and sulphur (and carbon, which is added separately) are regarded as macroelements. These elements comprise at least 0.1% of the dry weight of plants.

## Microelements

Elements are required in trace amounts for plant growth and development. Manganese, iodine, copper, cobalt, boron, molybdenum, iron and zinc are regarded as microelements, although other elements like aluminium and nickel are frequently found in some formulations.

## Iron Source

Iron is usually added in the medium as iron sulphate, although iron citrate can also be used. Ethylenediaminet-etraacetic acid (EDTA) is usually used in conjunction with the iron sulphate. The EDTA complexes with the iron so as to allow the slow and continuous release of iron into the medium. Uncomplexed iron can precipitate out of the medium as ferric oxide.

## Organic Supplements

These include vitamins and amino acids. Two vitamins, i.e., thiamine (vitamin B1) and myoinositol (a vitamin B) are essential for the culture of plant cells in vitro. However, other vitamins are often added to for historical reasons. The most commonly used amino acid is glycine. However, arginine, asparagine, aspartic acid, alanine, glutamic acid, glutamine and proline are also used. Amino acids provide a source of reduced nitrogen and, like ammonium ions, uptake causes acidification of the medium. Casein hydrolysate can be used as a source of a mixture of amino acids.

## Carbon Source

The most commonly used carbon source is sucrose. It is readily assimilated and relatively stable. Other carbohydrates like glucose, maltose, galactose and sorbitol can also be used and may prove better than sucrose in specialized circumstances.

## Gelling Agents

Plant tissue culture media can be used in either liquid or 'solid' forms, depending on the type of culture being grown. Agar, produced from seaweed, is the most common type of gelling agent, and is ideal for routine applications. For more demanding applications, a range of purer gelling agents are available. Purified agar or agarose can be used, as can a variety of gellan gums. *Plant growth regulators*: Specific media manipulations can be used to direct the development of plant cells in culture due to plasticity and totipotency. Plant growth regulators are the critical media components in determining the developmental pathway of the plant cells.

*There are five main classes of plant growth regulator used in plant cell culture, namely:*

1. Auxins
2. Cytokinins
3. Gibberellins
4. Abscisic acid
5. Ethylene.

## Auxins

Auxins promote both cell division and cell growth. IAA (indole-3-acetic acid) is the most important naturally occurring auxin but its use in plant tissue culture media is limited because it is unstable to both heat and light. 2,4-Dichlorophenoxyacetic acid (2,4-D) is the most commonly used auxin and is extremely effective in most circumstances.

## Cytokinins

Cytokinins promote cell division. Of the naturally occurring cytokinins, only zeatin and 2iP (2-isopentyl adenine have some use in plant tissue culture media. The synthetic analogues, kinetin and BAP (benzylaminopurine), are used more frequently. Non-purine-based chemicals, such as substituted phenylureas, are also used as cytokinins in plant tissue culture media.

## Gibberellins

Gibberellins are involved in regulating cell elongation, in determining plant height and fruit-set. Only a few of the gibberellins like GA3 are used in plant tissue culture media.

## Abscisic Acid

It is used in plant tissue culture to promote distinct developmental pathways such as somatic embryogenesis. Abscisic acid (ABA) inhibits cell division.

## Ethylene

Ethylene is associated with controlling fruit ripening in climacteric fruits, and its use in plant tissue culture is not widespread. Some plant cell cultures produce ethylene, which, if it builds up sufficiently, can inhibit the growth and development of the culture.

## Antibiotics

Antibiotics are substances produced by certain microorganisms that suppress the growth of other microorganisms and eventually destroy them. *Their applications include*:

- Suppresses bacterial infections in plant cell and tissue culture.
- Suppresses mould and yeast infections in cell cultures.
- Eliminates Agrobacterium species after the transformation of plant tissue.

*These antibiotics can be divided into different classes on the basis of chemical structure and their mechanism of action*:

- Inhibitors of Bacterial Cell Wall Synthesis.
  e.g. â-lactam antibiotics, Penicillins and Cephalosporins.
- Antibiotics that affect Cell Membrane permeability.
  - Antibacterial e.g. Colistin Sulphate, Polymixin B Sulphate, Gramicidin
  - Antifungal e.g. Amphotericin B, Nystatin, Pimaricin
- Bacteriostatic Inhibitors of Protein Antibiotics that affect the function of 30 S or 50 S ribosomal subunits to cause a reversible inhibition of protein synthesis. e.g. Chloramphenicol, Chlortetracycline HCl, Clindamycin HCl, Doxycycline HCl, Erythromycin, Lincomycin HCl, Oxytetracycline HCl, Spectinomycin sulphate, Tetracycline HCl, Tylosin tartrate, Lincomycin HCl
- Bactericide Inhibitors of Protein Synthesis Antibiotics that bind to the 30 S ribosomal subunit and alter protein synthesis which eventually leads to cell death. This group includes:
  - *Aminoglycosides*: Apramycin, Butirosine, Gentamicin, Kanamycin, Neomycin, Streptomycin, Tobramycin.
  - *Inhibitors of Nucleic Acid Metabolism*: e.g. Rifampicin, Mitomycin C and Nalidixic acid.
  - Antimetabolites: Antibiotics, which block specific metabolic steps that are essential to microorganisms e.g.
  - Nucleic Acid Analogs, which inhibit enzymes essential for DNA synthesis. e.g. 5-Fluorouracil, Mercaptopurine

## Preparation of Plant Tissue Culture Medium

- Measure approximately 90% of the required volume of the deionized-distilled water in a flask/container of double the size of the required volume.

- Add the dehydrated medium into the water and stir to dissolve the medium completely. Gentle heating of the solution may be required to bring powder into solution.
- Add desired heat stable supplements to the medium solution.
- Add additional deionized-distilled water to the medium solution to obtain the final required volume.
- Set the desired pH with NaOH or HCl.
- Dispense the medium into culture vessels.
- Sterilize the medium by autoclaving at 15 psi (121°C) for appropriate time period. Higher temperature may result in poor cell growth.
- Add heat labile supplements after autoclaving.

## DEMONSTRATION OF TISSUE CULTURE FOR TEACHING

The starting point for all tissue cultures is plant tissue, called an explant. It can be initiated from any part of a plant - root, stem, petiole, leaf or flower - although the success of any one of these varies between species. It is essential that the surface of the explant is sterilised to remove all microbial contamination.

Plant cell division is slow compared to the growth of bacteria and fungi, and even minor contaminants will easily over-grow the plant tissue culture.

The explant is then incubated on a sterile nutrient medium to initiate the tissue culture. The composition of the growth medium is designed to both sustain the plant cells, encourage cell division, and control development of either an undifferentiated cell mass, or particular plant organs.

The concentration of the growth regulators in the medium, namely auxin and cytokinin, seems to be the critical factor for determining whether a tissue culture is initiated, and how it subsequently develops. The explant should initially form a callus, from which it is possible to generate multiple embryos and then shoots, forming the basis for plant regeneration and thus the technology of micropropagation. The first stage of tissue culture initiation is vital for information on what combination of media components will give a friable, fast-growing callus, or a green chlorophyllous callus, or embryo, root or shoot formation.

There is at present no way to predict the exact growth medium, and growth protocol, to generate a particular type of callus. These characteristics have to be determined through a carefully designed and observed experiment for each new plant species, and frequently also for each new variety of the species which is taken into tissue culture. The basis of the experiment will be media and protocols that give the desired effect in other plant species, and experience.

## The Demonstration

The strategy for designing a medium to initiate tissue culture, showing how growth regulators and other factors modulate development, can be demonstrated using the African Violet, a popular house plant. Leaf sections are the source of explants.

This demonstration is regularly carried out by a student class, and gives reliable results. Sterile supplies are provided from central facilities, and provision of sterile working areas (for example, in laminar flow hoods) is an advantage, although cultures can be initiated in an open laboratory with careful aseptic technique. The standard precautions used during any laboratory work involving chemicals or microbes should be adopted. If you are in any doubt about safety hazards associated with this demonstration, you should consult your local safety adviser.

## Step 1: Selection of the Leaves

Leaves are cut from healthy plants, leaving a short length of petiole attached. They should be selected to each yield several explants of leaf squares with approximately 1 cm sides. The youngest and oldest leaves should be avoided.

Wash the dust off the leaves in a beaker of distilled water, holding the leaf stalk with forceps.

## Step 2: Surface Sterilisation and Preparation of the Explants

This part of the procedure should be carried out in a sterile working area, or with meticulous aseptic technique.

The leaf, with the petiole still attached, should be immersed in 70 per cent ethanol for 30 seconds, then transferred to a sterile petri dish. Sterile scissors and forceps are then used to cut squares from the leaf as explants, each with approximately 1 cm sides.

The explants are transferred into a 10 per cent hypochlorite bleach solution for 5 minutes, gently agitating once or twice during this time. They are then washed free of bleach by immersing in four successive beakers of sterile distilled water, leaving them for 2-3 minutes in each.

Three explants are placed on each petri dish of growth medium, with the upper epidermis pressed gently against the surface of the agar to make good contact. The petri dishes are sealed with plastic film to prevent moisture loss, and incubated at 25°C in 16h light/8h dark.

### Step 3: Assessment of Tissue Culture Development

The explants are incubated for 4 - 6 weeks, and inspected at weekly or fortnightly intervals. The growth of obvious bacterial or fungal colonies indicates contamination, and data from such cultures is obviously suspect. The development of dark brown tissue cultures can also be a consequence of contamination.The media used in the demonstration are designed to show the effects of auxin, cytokinin, sucrose and mineral salts on development. The media were based on the well-known Murashige and Skoog inorganic medium.

### Typical Results

Absence of sucrose inhibits this production. Shoot production is also limited on the low sucrose concentration, but comparable with the control at high sucrose.At zero and low levels of cytokinin, callus forms where the leaf surface is in contact with the medium, while at high levels, shoot formation is stimulated.At zero and low levels of auxins there is a stimulus to shoot formation, but at high concentrations, large numbers of roots are formed.At low and zero levels of MS salts, there is no growth at all.These very obvious variations demonstrate the importance of a carbon and inorganic salt source for plant growth, as well as the effect of the auxin:cytokinin ration on the control of plant development.

## THEORY OF TISSUE CULTURE

The theoretical basis for plant tissue culture was proposed by Gottlieb Haberlandt, German Academy of science in 1902 on his experiments on the culture of single cell. The first true cultures were obtained by Gautheret from cambial tissue of Acer pseudoplatanus.

The term plant tissue culture (Micro propagation) is generally used for the aseptic culture of cells, tissues, organs and their components under defined chemical and physical conditions *in vitro*. The basic concept of the plant body can be dissected into smaller part termed as "explants" and any explants can be developed into a whole plant. It is a central innovative areas of applied plant science, including agriculture and plant biotechnology.

This technique is effective because almost all the plants cell are totipotent; In each cell possesses the genetic information and cellular machinery necessary to generate the whole organism. Since, this technique can be used to produce a higher number of plants that are genetically similar to a parent plant as well as to another. Two concepts, plasticity and totipotency, are the central processes to understand the regeneration and plant cell culture. Plants, due to its longer life span and sessile nature, have developed a greater ability to overcome the extreme conditions.

Most of the processes include in plant development and the growth, adapt to environmental conditions. When the plant cells and tissues are cultured *in vitro,* most of them are generally exhibit a very high degree of plasticity, which allows one type of organ or tissue to be initiated from another type. Like this way, the whole plant can be subsequently regenerated. These maintenance of genetic potential is called totipotency. The plant tissue culture medium is an artificial nutrient supplement of organic and inorganic nutrients used for cultivation of plant tissue media.

The appropriate composition of the medium largely determines the success of the culture.

The culture media used for the *in vitro* cultivation of the plant cells are composed of three basic components.

- Essential elements (normal ions) supplied as a complex mixture of salts.
- An organic supplements providing vitamins and amino acids.
- A source of fixed carbon which is usually supplied as sucrose.

When cultured in an appropriate medium having auxin and cytokinin, explants will give rise to an unorganised, growing and dividing mass of cells called callus. Callus cultures are initiated from a small part of an organ or tissue segment called the explants on a growth supporting solidified nutrient medium under sterile conditions.

Any part of the plant organ or tissues may be used as the explants. At the time of callus formation, there is some degree of dedifferentiation happens both in morphology and metabolism. One of the major consequences of this dedifferentiation is that most plant cultures lose their ability to perform photosynthesis. The necessitates of the addition of other components such as carbon and vitamins source to the culture media, in addition to the unusual mineral nutrients.

## Morphology of callus

Callus varies considerably in appearance and texture, ranging from hard nodular cell masses to friable soft ones. They maybe white or creamish, orange, green either in whole or part as a result of chloroplast development. The shape of individual cells within the callus mass ranges from the near spherical or markedly elongated. A typical unorganised plant callus initiated from a new explants or piece of previously initiated calli has three stages of development.

1. The induction of cell division.
2. A period of active cell division during which differentiated cells lose specialised features they may have acquired and become dedifferentiated. Cell division usually occurs in the outer layer of the explants.
3. Period when cell division slows down on ceases and when within the callus, there is increasing cellular differentiate.

Callus culturing is performed in the dark while light can be encourage the differentiation of the callus. At the time of long term culture, the culture may loss the requirement for cytokinin and auxins. Manipulation of the auxins to cytokinin ratio in the medium can leads to the development of shoots, roots or somatic embryos from which the plant can be subsequently produced.

*Callus culture is useful for many purposes:*

- Callus is the starting material for the suspension culture which cells are separated.
- It helps in the production of secondary plant products.
- It is useful for the synthesis of starting compounds that are subsequently modified to yield the desired product.
- It is the starting materials for vegetative propagation of plants.

Based on the availability of the various *invitro* techniques, the dramatic increase in their application to various problems in basic biology, agriculture, horticulture, and forestry.

*The applications can divide conveniently into six broad areas:*

1. Cell behaviour.
2. Plant modification.
3. Germplasm storage and pathogen –free plants.
4. Clonal propagation.
5. Product formation.
6. Improved varieties.

## PATHWAYS OF CULTURED CELLS AND TISSUES

The cultured cells and tissue can take several pathways to produce a complete plant. Among these, the pathways that lead to the production of true-to-type plants in large numbers are the popular and preferred ones for commercial multiplication. The following terms have been used to describe various pathways of cells and tissue in culture.

## Regeneration and Organogenesis

In this pathway, groups of cells of the apical meristem in the shoot apex, axillary buds, root tips, and floral buds are stimulated to differentiate and grow into shoots and ultimately into complete plants. In many cases, the axillary buds formed in the culture undergo repetitive proliferation, and produce large number of tiny plants.

The plants are then separated from each other and rooted either in the next stages of micropropagation or in vivo (in trays, small pots or beds in glasshouse or plastic tunnel under relatively high humidity). The explants cultured on relatively high amounts of auxin (*e.g.* (2,4-D, 2,4-dichlorophenoxyacetic acid) form an unorganized mass of cells, called callus. The callus can be further sub-cultured and multiplied.

The callus shaken in a liquid medium produces cell suspension, which can be subcultured and multiplied into more liquid cultures. The cell suspensions form cell clumps, which eventually form calli and give rise to plants through organogenesis or somatic embryogenesis. In some cases, explants *e.g.* leaf-discs and epidermal tissue can also generate plants by direct organogenesis and somatic embryogenesis without intervening callus formation, *e.g.* in orchardgrass. Dactylis glomerata L.. In organogenesis the cultured plant cells and cell clumps (callus) and mature differentiated cells (microspores, ovules) and tissues (leaf discs, inter-nodal segments) are induced to differentiate into complete plants to form shoot buds and eventually shoots, and rooted to form complete plants.

## Somatic Embryogenesis

In this pathway, cells or callus cultures on solid media or in suspension cultures form embryo-like structures called somatic embryos, which on germination produce complete plants. The primary somatic embryos are also capable of producing more embryos through secondary somatic embryogenesis. Although, somatic embryogenesis has been demonstrated in a very large number of plants and trees, the use of somatic embryos in large-scale commercial production has been restricted to only a few plants, such as carrot, date palm, and a few forest trees.

Somatic embryos are produced as adventitious structures directly on explants of zygotic embryos, from callus and suspension cultures. Somatic embryos and synthetic seeds (embryos encapsulated in artificial endosperm) hold potential for large-scale clonal propagation of superior genotypes of heterogeneous plants.

They have also been used in commercial plant production and for the multiplication of parental genotypes in large-scale hybrid seed production. In many species, somatic embryos are morphologically similar to the zygotic embryos, although some biochemical, physiological and anatomical differences have been documented.

The synthetic auxin, 2,4-D is commonly used for embryo induction. In many angiosperms, *e.g.*, carrot and alfalfa, subculture of cells from 2,4-D containing medium to auxin-free medium is sufficient to induce somatic embryogenesis.

The process can be enhanced with the application of osmotic stress, manipulation of medium nutrients, and reducing humidity. Selection of embryogenic cell lines has also been successfully used. For example, selection for unique morphotypes in grapevine cultures allows production of high quality embryos with predicable frequency.

A major problem in large-scale production of somatic embryos is culture synchronization. This is achieved through selecting cells or pre-embryonic cell clusters of certain size, and manipulation of light and temperature, temporary starvation or by adding cell cycle synchronizing chemicals to the medium. Cytokinins seem to play a key role in cell cycle synchronization and embryo induction, proliferation and differentiation. Abscisic acid is crucial in all the stages of somatic development, maturation and hardening.

## Synthetic Seeds

The concept of production and utilization of synthetic seeds (somatic embryo as substitutes for true seeds) was first suggested by Murashige in 1977. Synthetic seeds can be produced either as coated or non-coated, desiccated somatic embryos or as embryos encapsulated in hydrated gel (usually calcium alginate).

Successful utilization of synthetic seeds as propagules of choice requires an efficient and reproducible production system and a high percentage of post-planting conversion into vigourous plants. Artificial coats and gel capsules containing nutrients, pesticides and beneficial organisms have long been thought as substitutes for seed coat and endosperm. However, this technology is still in the developmental stage, and currently cannot compete with the other methods of commercial plant propagation.

## Process of Micropropagation

The process of plant micro-propagation aims to produce clones (true copies of a plant in large numbers).

*The process is usually divided into the following stages:*

- *Stage* 0 –
  Pre-propagation step or selection and pre-treatment of suitable plants.
- *Stage I* –
  Initiation of explants-surface sterilization, establishment of mother explants.
- *Stage II* – subculture for multiplication/proliferation of explants.
- *Stage III* – shooting and rooting of the explants.
- *Stage IV* – Weaning/hardening.

These stages are universally applicable in large-scale multiplication of plants. The individual plant species, varieties and clones require specific modification of the growth media, weaning and hardening conditions.

A rule of the thumb is to propagate plants under conditions as natural or similar to those in which the plants will be ultimately grown ex-vitro. For example, if a chrysanthemum variety is to be grown under long day-length for flower production, it is better to multiply the material under long-day length at stages III and IV. There is a wide option to undertake production of plant material up to a limited number of stages. For example, many commercial tissue culture companies undertake production up to Stage III, and leave the remaining stages to others.

## Pre-Propagation Stage

The pre-propagation stage (also called stage 0) requires proper maintenance of the mother plants in the greenhouse under disease- and insect-free conditions with minimal dust. Clean enclosed areas, glasshouses, plastic tunnels, and net-covered tunnels, provide high quality explant source plants with minimal infection. Collection of plant material for clonal propagation should be done after appropriate pretreatment of the mother plants with fungicides and pesticides to minimize contamination in the in vitro cultures. This improves growth and multiplication rates of in vitro cultures.

The control of contamination begins with the pretreatment of the donor plants. They may be prescreened for diseases, isolated and treated to reduce contamination. The explants are then brought to the production facility, surface sterilized and introduced into culture. They may at this stage be treated with antibiotics and fungicides as well as anti-microbial formulations, such as PPM. The explants are then culture indexed for contamination by standard microbiological techniques, which are occasionally supplemented with tests based on molecular biology or other techniques.

## Stage I

This stage refers to the inoculation of the explants on sterile medium to initiate aseptic culture. Initiation of explants is the very first step in micropropagation. A good clean explant, once established in an aseptic condition, can be multiplied several times; hence, explant initiation in an aseptic condition should be regarded as a critical step in micropropagation. More than often, explants fail to establish and grow, not due to the lack of a suitable medium but because of contamination. The explants are transferred to in vitro environment, free from microbial contaminants. The process requires excision of tiny plant pieces and their surface sterilization with chemicals such as sodium hypochlorite, ethyl alcohol and repeated washing with sterile distilled water before and after treatment with chemicals. After a short period of culture, usually 3 to 5 days, the contaminated explants are discarded.

The surviving explants showing growth are maintained and used for further subculture. In herbaceous plants *e.g.* potato, chrysanthemum, carnation, streptocarpus, strawberry, and African violet; the explant sources are meristems, apical- and axillary buds, young seedlings, developing young leaves and petioles, and unopened floral buds. The following low cost options can be adapted to initiate explants:

## Sterile Instrument Technique

This method assumes that most of the deep-seated meristems and those covered by leaves or other integuments (*e.g.* floral bracts) are sterile. In this procedure, the explant is washed with sterile water, rinsed in ethanol, and instruments are sterilized every time they touch the surface of the explant, and the explant is moved to a new location on the dissection stage.

## Surface Sterilization Technique

This is by far the most commonly used method. The explants are washed in sterile water, rinsed in ethanol, and surface sterilization is achieved by using chemicals with chlorine base. Calcium or sodium hypochlorite based solutions, 1–3 per cent (v/v) are usually used for soft herbaceous materials. A cheap and ready-made sterilant is 5-7 per cent solution of 'Domestos'- a toilet disinfectant which contains 10.5 per cent v/v sodium hypochlorite, 0.3 per cent sodium carbonate, 10.0 per cent sodium chloride and 0.5 per cent (w/v) sodium hydroxide and a patented thickener). The explants are washed in sterile distilled water before and after sterilization. Other surface sterilants used include mercuric chloride (avoid its use as far as possible, since it is highly toxic), hydrogen peroxide, and potassium permanganate.

## For Soft Tissues

- Wash explants from perennial plants for 1-2 hr in tap water. Eliminate this step for material from glasshouse grown plants.
- Wash in sterile distilled water three to four times for 5 to 10 minutes each.
- Dip in 95 per cent ethanol for 3 to 5 seconds.
- Wash once again with sterile distilled water for 5 minutes.
- Surface-sterilize in 5 per cent 'Domestos' (v/v) for 20-25 minutes.
- Wash with sterile distilled water three times for 10 minutes each.
- Drain water droplets by placing on pre-sterilized blotting paper.
- Transfer explants singly to the medium.

N.B.: Sterilize forceps each time to transfer explants to avoid cross-contamination.

*For woody stems (e.g. roses, hardy shrubs, and trees):*

- Collect stems, shoots, buds and store at 5 oC till needed.
- Rinse in ethanol for 3 to 5 seconds.
- Rinse in 1- per cent sodium hypochlorite (20 per cent bleach) for 10 minutes.
- Place lower parts of stems in flasks in 2 per cent sucrose and 200 PPM 8-hydroxyquinoline citrate at 23+2 oC. For items collected in September/ October, add 50-PPM GA3.

*After that l0 PPM GA will help break the dormancy:*

- Re-cut the bottom of stem and replace the solution after 2 days.
- Excise the softwood from the developed shoots and use material for explants or for rooting.
- Surface-sterilize as in the above protocol.

Do not forget to sterilize forceps and scalpel every time for the transfer of explants to fresh solutions. Use sterile containers in the protocol of surface sterilization. If explants become brown or pale at the end of the protocol, reduce the strength of 'Domestos' to 2.5 per cent. Alternatively, dip explants in 10 per cent 'Domestos' for 2 minutes and then proceed to surface sterilize with 3-5 per cent 'Domestos' for 20 minutes. If basal contamination is observed after 2-3 days of culture, explants can sometimes be rescued by removing the basal end by making a single cut with a sharp scalpel and re-culturing on fresh medium.

### Stage II

Stage II is the propagation phase in which the explants are cultured on the appropriate media for multiplication of shoots. The primary goal is to achieve propagation without losing the genetic stability.

Repeated culture of axillary and adventitious shoots, cutting with nodes, somatic embryos and other organs from Stage I leads to multiplication of propagules in large numbers. The propagules produced at this stage can be further used for multiplication by their repeated culture. Sometimes it is necessary to subculture the in vitro derived shoots onto different media for elongation.

### Stage III

The in vitro shoots obtained at Stage II are rooted to produce complete plants. If the proliferated material consists of bud-like structures (*e.g.* orchids) or clumps of shoots (banana, pineapple), they should be separated after rooting and not before. Many plants (*e.g.* banana, pineapple, roses, potato, chrysanthemum, strawberry, mint, several grasses and many more) can be rooted on half-strength-MS medium without any growth-regulators.

Good sturdy well-rooted plants are essential for high survival during weaning and later transfer to soil. This stage is labour intensive and expensive. The process of in vitro rooting has been estimated to account for approximately 35–75 per cent of the total cost of production. Efforts should be made to combine rooting and acclimatization stages.

### Stage IV

At this stage, the in vitro micropropagated plants are weaned and hardened. This is the final stage of the tissue culture operation after which the micropropagated plantlets are ready for transfer to the greenhouse. Steps are taken to grow individual plantlets capable of carrying out photosynthesis. The hardening of the tissue-cultured plantlets is done gradually from high to low humidity and from low light intensity to high intensity conditions. If grown on solid medium, most of the agar can be removed gently by rinsing with water.

Plants can be left in shade for 3 to 6 days where diffused natural light conditions them to the new environment. The plants are then transferred to an appropriate substrate (sand, peat, compost, etc.), and gradually hardened.

Low-cost options include the use of plastic domes or tunnels, which reduces the natural light intensity and maintains high relative humidity during the hardening process. If the plants are still joined together after rooting, these should be planted as bunches in the soil and separated after 6 to 8 weeks of growth.

## Delivery to the Growers

The delivery of the rooted and hardened small micropropagated plants to growers and market requires extra care. In some cases, plant losses can occur during shipment and handling by growers. This is particularly true when the plants are not fully hardened and rooted or not grown for sufficient duration after transfer to soil. Growers should be given clear instructions how to handle the material provided. Apart from the economic loss, poor survival of planted material erodes the confidence of growers in the technology. The transfer of individual plants to soil in black plastic or polythene bags is widely used as a low-cost option to provide fully-grown banana plants directly to farmers in many developing countries.

# 4

# Agricultural Investments in Plant Biotechnology

## AGRICULTURAL BIOTECHNOLOGY REGULATION

For traditionally bred plants, regulators rely on plant breeders to conduct appropriate safety testing and to be the first line of defense against genetic alterations that might prove dangerous.

The same should be true for agricultural products developed using biotechnology. There is no reason for regulators to treat biotech plants any differently than traditionally bred plants, particularly given the fact that biotechnology provides greater control over gene manipulation. Despite fears about gene manipulation, traditional crossbreeding has altered plant genes and improved the human diet for the past 30 years. In fact, people worldwide safely consume these plants every day.

Examples include wheat, corn, potatoes, tomatoes, and countless other staples of the American diet. In 1984, the biotechnology industry began experimenting with "gene-spliced plants" — the more advanced approach to gene manipulation that we now call "biotechnology."

### Department of Agriculture

USDA regulates the release of all genetically engineered agricultural plants under statutes giving USDA's Animal and Plant Health Inspection Service (APHIS) the authority to regulate plants that may be or may become weeds or other nuisances — what the statutes call "plant pests." Although the rules apply in a general sense to novel or exotic varieties of both gene-spliced and conventional plants, APHIS typically only requires field testing of conventional plants that are new to a particular U.S. ecosystem (transplanted from another continent, for example).

However, all genetically engineered agricultural plants face a higher regulatory bar — making it more difficult to expand the use of biotechnology. Genetically engineered crops must be field tested under APHIS regulations prior to commercialization. For example, a new variety of corn produced with conventional hybridization requires no government-mandated field testing, but all new varieties of genetically engineered corn do, even though there is no logical reason for the regulatory disparity. For most genetically engineered plants, APHIS requires the company producing the plants to submit notice detailing the gene or genes that have been inserted, where the plant will be tested, and other relevant characteristics of the plant prior to receiving permission to conduct the field trials. Once the company completes field testing, APHIS reviews the results and makes a determination on whether or not the product should be "deregulated" and can be released into the market.

## Regulatory Scheme

At the time, the White House Office of Science and Technology Policy began crafting a framework wherein existing federal agencies would regulate genetically engineered organisms on the basis of their characteristics, not the method of production — thus wisely deferring to the scientific community's judgment that regulation ought to address the products, not the process of biotechnology.

Engineered organisms would not require extra scrutiny simply because genetic engineering produced them (the process). Instead, they would be subject to heightened scrutiny only if the individual organisms expressed characteristics that posed some conceptually heightened risk (the product).

The federal government divided regulatory jurisdiction among agencies already involved in agricultural, food, and environmental regulation. These include the United States Department of Agriculture (USDA), the Environmental Protection Agency (EPA), and the Food and Drug Administration (FDA). While each of these agencies considers the characteristics of individual products in their regulation, only FDA followed the general scientific thinking that genetically engineered and non-genetically engineered products should be regulated similarly. Both USDA and EPA automatically subject all genetically engineered plants as a class to premarket approval requirements not ordinarily applied to conventionally bred plants.

## Environmental Protection Agency

EPA regulates plants that are modified to produce substances that act like pesticides: that is, substances used by a plant to protect itself from pests, such as insects, viruses, and fungi. EPA's proposed rule for these "plant-incorporated protectants" is not yet finalized. In the interim, however, EPA has regulated such plants by applying its proposed guidelines, which are functionally similar to rules for the registration of synthetic chemical pesticides. Again, biotech products face higher regulatory hurdles, even though plant-incorporated protectants developed through conventional breeding are exempt from these requirements.

## Health and Human Services: Food and Drug Administration

FDA is responsible for ensuring that food items, including foods derived from genetically modified plants, are safe to eat. Following the general regulatory framework that emphasizes product regulation rather than process regulation, FDA rightly does not treat foods derived from genetically engineered plants to be inherently unsafe.

Food producers are not required to seek premarket approval from FDA unless there is a substantive reason to believe that the novel trait (or traits) in the food pose a safety question. The initial determination of safety is left to the producer, but FDA has encouraged producers to consult with agency scientists prior to marketing a food produced with biotechnology to ensure that the appropriate determination is made. Recently, FDA published a proposed rule that would require producers to notify the agency at least 120 days before marketing, which may be a signal that FDA could abandon its more reasonable approach in the future.

In addition, FDA does not require labelling of foods derived from biotechnology unless the genetic insertions so alter the food that the common name no longer describes it adequately. Examples of this phenomenon would include such alterations of the food product that would raise a safety issue or change the product's nutritional composition or its storage or preparation characteristics.

## International Trade

While U.S. consumers do not appear to be strongly opposed to biotech foods (they, in fact, seem rather indifferent), a strong anti-biotechnology movement has arisen in several European countries over the last few years. The European Union (EU) has established strong restrictions on the commercial planting of genetically engineered plants, and European food processors and retailers are reluctant to import harvested agricultural products derived from biotechnology — thus jeopardizing the marketability of U.S. commodity grain exports. And the EU is now negotiating for strong restrictions on agricultural biotech products in important international agreements governing food safety and environmental protection. Very strong restrictions were included in the Biosafety Protocol, finalized in January 2000.

Perhaps, more important, though, are the Codex Alimentarius Commission standards for food safety. Without convincing scientific evidence that genetically engineered crop plants pose a heightened environmental or human health risk, restrictions on agricultural biotech imports could be challenged under the General Agreement on Tariffs and Trade and the World Trade Organization. Thus the EU has resorted to justifying its decisions on the basis of a risk-management philosophy known as the "precautionary principle," and the EU advocates inclusion of the precautionary principle in international environmental and food safety agreements (such as the Codex Alimentarius), as well as within GATT itself. Its inclusion would give nations greater latitude in restricting imports of U.S. agricultural products.

No single definition exists for the precautionary principle, but its general meaning is that when an activity raises threats of harm to human health or the environment, regulatory measures should be taken to prevent or restrict the activity even if the risks have not been demonstrated scientifically. Although the EU asserts that the precautionary principle is an unbiased risk-management philosophy, critics have noted that its lack of definition and evidentiary standards makes it all too easy to abuse for the purpose of masking trade protectionism, and that its very approach to risk management is inherently flawed and may, in fact, increase net risk.. Nevertheless, the precautionary principle has already been included in several international environmental treaties, including the Convention on Biological Diversity and the Biosafety Protocol.

## Labelling Issues

Some activists, however, argue that the government should mandate the labelling of all genetically engineered foods. They assert that consumers have a "right to know" how their foods have been altered, and that a mandatory label would best allow consumers to choose between genetically engineered and conventional foods. Biotechnology advocates have argued against mandatory labelling because such requirements raise food costs— something that mostly harms lower-income Americans and people on fixed budgets. Perhaps more important, while biotech products are not substantially different from other products, special labels would likely make consumers think these products were more dangerous. Hence, rather than serving educational or "right to know" purposes, such labels promise to simply confuse consumers.

A government-mandated label on all genetically engineered foods also would raise important First Amendment free speech issues. In 1996, the U.S. Court of Appeals, in *International Dairy Foods Association et al. v. Amestoy*, ruled unconstitutional a Vermont statute requiring the labelling of dairy products derived from cows treated with a genetically engineered growth hormone, noting that food labelling cannot be mandated simply because some people would like to have the information. "Absent ... some indication that this information bears on a reasonable concern for human health or safety or some other sufficiently substantial governmental concern, the manufacturers cannot be compelled to disclose it."

In other words, to be constitutional, labelling mandates must be based in science and confined to requiring disclosure of information that is relevant to health or nutrition. Furthermore, consumers need not rely on mandatory labelling of biotech foods to truly have a choice. Real-world examples show that market forces are fully capable of supplying information about process attributes (including kosher and organic production standards) that consumers truly demand. The same can be said about nonbiotech foods, and the FDA recently published proposed guidelines to assist producers in voluntarily labelling both genetically engineered and nongenetically engineered foods. Additionally, the USDA's newly published rule for organic certification necessarily excludes biotech products from organic food production.

Consequently, consumers wishing to purchase nonbiotech foods need only look for certified organic products.

Even as underdeveloped nations clamor for biotechnology applications, and as countries like China continue to experiment with and use agricultural biotechnology, opponents of agricultural biotechnology in the West, particularly Europe, attack it as an unnatural process that will destroy the world, not better it. They argue that biotechnology should be heavily regulated, if not banned. Already, however, genetically engineered plants are subject to strict regulatory oversight that is equal to or greater than that advocated by the vast majority of scientific specialists.

Additional regulation will slow down research and development of genetically engineered crops, keep beneficial products off the market, and raise the cost of products that do make it to consumers. Furthermore, the inclusion of similar restrictions — or inclusion of the precautionary principle — in international agreements will greatly impact the international trade of agricultural goods and delay their introduction into the marketplace. Each of these problems could prevent this technology's benefits from being introduced to industrialized nations and, more important, the developing world.

## MODERN AGRICULTURAL BIOTECHNOLOGY

In 1973, Cohen and Boyer transferred a gene from one organism into another. In 1982, the first biotech plant, an antibiotic resistant tobacco, was developed. In January 1983, at a meeting of genetic researchers in Miami, three different teams reported success in using Agrobacterium tumefaciens, a bacterium, to carry new genes into plant cells, heralding the dawn of modern agricultural biotechnology. Agrobacterium tumefaciens, described as a "natural genetic engineer", splices its own genes into host plant cells. This pathogenic bacterium was now converted into a pack mule, to carry new, foreign genes into plant cells, minus the disease and this became the most common means of producing Genetically Engineered Organisms (GEOs).

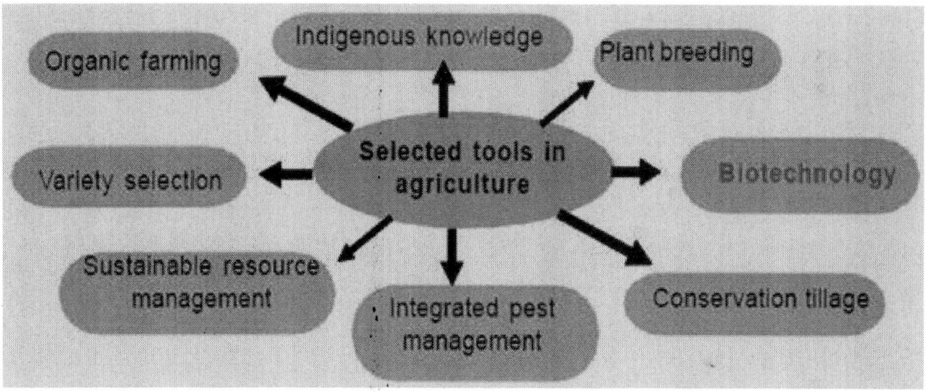


*Figure: Modern Agricultural Biotechnology.*

Field tests for GE crops resistant to pests and pathogens were first conducted in the US, in 1985. A co-coordinated framework for the regulation of GEOs was established and the first GE tobacco was released, in 1986 . The US Department of Agriculture published guidelines for field trials of GE crops in 1991.

On approval from the Food and Drug Administration, Flavor Saver, the first GE tomato, with a longer shelf life, was on the US markets in 1994. During 1995-96, GE soybean, corn and cotton were approved for commercialization, in the US.

A number of GE crops, developed for pest, pathogen and herbicide resistance, are now commercially cultivated in several countries. Rice with pro-vitamin A, higher iron content, or human milk proteins and potatoes with high protein content are in various stages of development. Tobacco plants producing functional human haemoglobin and bacteria that produce human insulin have been developed, as well as plants with vaccines against rabies and other viral diseases. Food grain crops that withstand drought and salinity are high priority research, and so are those for high yield. A GE tobacco plant detoxifies soils contaminated by explosive residues, providing solution to a frustrating environmental problem in countries ravaged by armed conflict.

Now more than 70 biotech agricultural crops that have been approved for use in North America, including varieties of soybeans, cotton, canola, corn, potatoes, squash, tomatoes and papaya. About six million farmers in some 17 countries now cultivate GE crop on about 125 million acres, a 30 fold increase over 1996. By end of the year 2002, six GE crops planted in the US (soybeans, corn, cotton, papaya, squash and canola) produced an additional four billion pounds of food and fibre on the same acreage, improved farm income by US $ 1.5 billion and reduced pesticide use by 46 million pounds. In 2003 in the US, 80 per cent of soybean acres will be planted with biotech varieties.

A number of activist groups vehemently attack GE technology and its products on grounds of safety to humans and the environment, and costs of technology transfer and its reach to the needy. Products of agricultural biotechnology bear the brunt of this ill-informed, unscientific and prejudiced onslaught much more than GE products related to health or industry.

In 2001, the European Community released results of a 15-year study, costing US $ 64 million, and involving more than 400 research teams and 81 projects. This report concluded that GE products pose no more risk to human health or the environment than conventional crops. So far, extensive and intensive research on the probable risks of GE technology has not brought out any adverse effects and none of the fears expressed by anti-tech activists were proved even marginally. Nevertheless, caution and examining issues case by case, is the watch word of technologists, who are aware of their responsibilities.

For a number of products, technology transfer is free of costs for developing countries, as for example Golden Rice, the rice with pro-vitamin A. It is the responsibility of the Governments of the respective countries to bear the costs of developing local varieties and to reach the products to the needy at an affordable cost. In India, three GE varieties of pest resistant cotton were approved for commercialization, a year ago. Approval for other GE varieties of cotton and GE mustard was deferred twice by the Genetic Engineering Approval Committee (GEAC), the highest authority on the issue.

In India, the level of public awareness of the realizable benefits and probable risks of GE products is abysmally low. The functioning of the GEAC leaves much to be desired. Taking advantage of this hazy situation, mixing up economic, social and political issues with science, and even using such vagaries of nature as the severe drought, several groups of vested interest have created mistrust, confusion and scare. Trashing a very promising technology this way does not augur well for the future of Indian agriculture. This results only in denying the benefits of technology to the farmers and consumers.

The next 20 years of plant biotechnology is expected to improve quality of life, through improved foods, pharmaceuticals, new industrial materials and a better environment. These benefits should reach the people of the developing world, who need them most. For India to benefit from these developments, it is urgent and essential that the public is educated on the realizable benefits and probable risks related to GE products. We also need to reorganize the mechanism of regulating GE products by providing for an expedient, transparent and responsible authority. The media have a very important role to play in this regard. Only an informed and reassured public can make viable choices. Every effort should be made to provide such an opportunity to the consumer.

## GENERAL BIOTECHNOLOGY INDICATORS

Since 1999, organizations such as the Organization for Economic Co-operation and Development (OECD) and the Organization of American States (OAS) began to compile statistics to describe the situation of the countries that are developing biotechnologies. A general view shows that countries in Latin America and the Caribbean (LAC) have significant gaps with countries of the North, especially in patterns of patenting and licensing technologies, and that the transferal of biotechnologies is of vital importance, especially those biotechnologies developed for the agricultural sector (Testa, 2002; Heisey et al. 2005; Thumm, 2005; Chan, 2006).

In LAC countries, measurements mainly include the analysis of the capacity of groups and research centers, their working fields, the number of enterprises and the market perspectives, whereas the statistics of OECD measure mainly the private investment, the generation of investor's capital in terms of new companies, money and employment, and the obtainment of patents among other indicators, which are not so representative for us due to the differences in economies and development systems.

An indicator to analyze the state of biotechnology in the countries is a function of the investment made for its developments and products. The

United States of America has a much higher investment tendency in relation to other countries. In 1997, the industry invested around US$ 7.300 millions, being 3.7% of the total investment for research and development (NSF, 2004), and as for 2001 it reported that investment of industry in research and development of biotechnology reached US$ 16.400 millions, representing 10% of the total investments in science and technology (U.S. Department of Commerce, 2003). European countries such as Germany, with more than US$ 1.000 millions (6.7% of total), United Kingdom with US$ 705 millions (7.8% of total) and France with US$ 560 millions having a 4.4% of the total of governmental investment in R&D, follow the list with the largest assigned budget for biotechnology according to statistics of OECD (Van Beuzekom, 2001).

The situation in countries members of the OECD reveals a striking difference between USA and EU. According to data provided in 2001 by the consultant company Ernst and Young, USA investment in R&D was € 11.400 million (2.7% of DGP) whilst EU investment was € 4.977 million (1,8% of DGP). There are 1,262 companies in USA totalling 162,000 employees and 6.7% of them are involved in research, in contrast to 1,570 companies in EU with 61,000 employees, 2.5% of them conducting research (Ernst and Young, 2001).

In the compendium of biotechnology statistics of OECD, differences can also be observed. The first one is the amount of investment in risk capitals in biotechnology, which goes up to US$ 1.182 millions in the USA, and US$ 687 millions for the EU during 1999. The bioengineering products reached that year up to 1% of the total exports of USA with a total amount of US$ 1.340 millions. Its import of products from OECD countries, especially from Belgium, France, Switzerland and The Netherlands were that same year of US$ 970 millions. Its most important commercial partners where most of the exports are sent to include: Belgium, Japan, Canada and Germany (Van Beuzekom, 2001).

This indicator allows analyzing the degree of technological delay resulting from the efforts of countries such as Colombia, where it represents approximately US$ 1.8 millions (constant Colombian pesos as in 2003. Exchange rate 1 US$ = $ 2.200 Col) and 3.7% of the total governmental investment of the National Biotechnology Program (NBP) made through Colciencias during 1997, and during the following years the tendency remained negative, reaching in 1999 almost US$ 780.000 and 4.8% (OCyT, 2004).

This situation somehow explains why Colombia reports 8 patents granted to research groups associated with the NBP (OCyT, 2004), whereas the USA already had more than 3.600 patents and Canada more than 500 during the year 2000 (Van Beuzekom, 2001). It could be said that the relation between the investment in biotechnology and the number of patents is more than US$ 2 million per patent in the USA, US$ 640.000 in Canada and US$ 22.000 in Colombia.

On the other hand, the development of biotechnology in LAC has been characterized by being a repetitive model where the research projects do not correspond to the real needs of production and food security, and has a low level of innovation and scientific creativity. Only 2% out of 30 scientific articles of high impact on biotechnology came from laboratories based within the region (Izquierdo and De La Riva, 2000), despite LAC possessing important biodiversity resources to develop innovative products (Roca et al. 2004).

## AGRICULTURAL BIOTECHNOLOGY SECTOR

Agricultural biotechnology sector in India, particularly its plant biotechnology segment, is uniquely poised for a major process of transformation. As the country's agricultural biotechnology sector made strides in the 1990s in r-DNA, transgenics and molecular marker assisted plant breeding process, the Government of India responded with a matching policy support and regulatory framework that was designed to render the path of progress in R&D, sustainable and bio-safe. To a large extent, developments in the policy front have been induced by a vibrant non-governmental sector that intensely intervened on the sensitivities of modern biotechnology. This chapter discusses analysis of the issues and proceeds to further consider the path of progress achieved by agricultural biotechnology companies in India in the area of commercialization of biotechnology products. The special focus of this chapter is on plant biotechnology. The biotechnology sector in India in 2002-03 is estimated to have attracted private investments to the tune of US$ 140 million. Infrastructure and R&D account for nearly 80 per cent of these investments.

During the period from 1986 to 2002 it is estimated that the Government of India invested an amount of US$ 275 million in the biotechnology sector. The Government of India has also extended venture capital support for biotechnology start-ups through the Technology Development Board and the New Millennium Indian Technology Industry Leadership Initiative (NMITLI). Meanwhile the States of Andhra Pradesh, Karnataka, Maharashtra, Gujarat, Kerala and Tamil Nadu have drawn up ambitious sub-national biotechnology policies, which promise fiscal and infrastructure support measures to prospective entrepreneurs. Some State Governments have launched their own venture funds for supporting novel start-up ventures.

To catalyse the process, the Government of India accorded priority to IPR protection. The Indian Patents Act of 1970 was amended to provide for product patents for agro-chemicals, drugs and pharmaceuticals and micro-organisms over a transitory period. A product patent regime would be in place in India by January 1, 2005. The new patent law extends protection term for inventions from 14 to 20 years. To provide for protection of IPRs in respect of new plant varieties, Parliament has enacted a *sui-generis* legislation in the shape of the Plant Varieties Protection and Farmers Rights (PVFR) Act in 2001. These developments have created favourable legal conditions for international partnerships in biotechnology R&D. By prioritizing FDI investments in biotechnology, the Government of India has sent appropriate signals to international investors. Further the fact that these changes have been in consonance with the WTO-TRIPs has lent greater creditability to these changes. Meanwhile, the Government has also brought in regulations over agricultural biotechnology. Apart from the biosafety regulations that form part of the Environment (Protection) Act 1986, the Indian Parliament has legislated the National Biodiversity Act in the year 2002, to provide for regulatory controls over access to biological resources in India.

## The Changing Structure Of The Agricultural Biotechnology Industry

The top ten pesticide and seed firms for 1997 and 1999. Longer-term time series data on the structure of the chemical and seed company are difficult to obtain. In one of the few published papers to provide this data, Ollinger and Fernandez-Cornejo (1998) provide Concentration Ratio data (CR4) for the United States pesticide industry over the period 1972-1989. During this period, the CR4 oscillates, moving from a high of 50% in 1973 to a low of 37% in 1982, and then rising to 48% in 1989.

Although detailed data are not available for the 1990s, there is evidence that the structure of the seed and chemical industry remains dynamic. The consolidation activity of the ten most active biotechnology firms. Mergers and acquisitions are a key component of the new business relationships that have emerged. As Kalaitzandonakes and Hayenga (2000) note, firm entry in the crop biotechnology industry peaked in the early 1980s, with production innovation continuing throughout the 1980s. Oehmke et al. (1999) also discuss the cyclical pattern in mergers in the biotechnology industry. Aggregate market figures mask the much higher concentration that exists in specific markets (Sexton, 2000). For example, in 1998, Monsanto and Pioneer-HiBred (now owned wholly by DuPont) controlled 15% and 39% of the US seed corn market, respectively. These two same companies controlled approximately 24% and 17%, respectively, of the purchased soybean seed market. For the US cotton market, two companies, Delta & Pine Land and Stoneville, had 71% and 16%, respectively, of the seed market (Kalaitzandonakes & Hayenga, 2000).

***Figure: Crops developed with agricultural biotechnology reduce the need for tillage***

The determination of the relevant market concentration is not always done on the basis of output markets. The Federal Trade Commission has used innovation competition to assess the impact of mergers. Examining competition in innovation is designed to focus attention on the impact of a merger on innovative activity. Using data on field trials underway each year, Brennan Pray, and Courtmanche (2000) calculate a CR4 ratio at the R&D stage. In 1988, the four largest firms had 87% of the field trials. The CR4 ratio declined to a low of 63% in 1995, then rose steadily over the next few years to reach 79% in 1998. This concentration at the R&D stage is matched by concentration in terms of the number of patents held. The top four firms held 41% of the corn patents (up to 1996), 53% of the soybean patents (up to 1997), 77% of the tomato patents (up to June 1997) and 38% of the Bt patents (up to 1998) (Brennan, Pray, & Courtmanche, 2000).

Concentrated markets do not necessarily imply the presence of market power. Baumol, Panzar and Willig (1992) stress that firms will not be able to exercise market power if the market is contestable. The key requirements for market contestability are: (a) Potential entrants must not be at a cost disadvantage to existing firms, and (b) entry and exit must be costless. For entry and exit to be costless, there must be no sunk costs. Sunk costs are expenditures that cannot be recouped once they are incurred—examples include expenditures made to obtain regulatory approval, expenditures on advertising, and expenditures on R&D. If there are no sunk costs, potential entrants can use a hit-and-run strategy in which they enter an industry, undercut the price of the incumbents, reap the profits and exit before the incumbents have time to retaliate. In anticipation of entrants acting in this manner, the incumbents forestall entry by keeping price at average cost. The consequence is that, even in an industry that is highly concentrated, prices can be kept at or near competitive levels.

However, if sunk costs are present, firms entering an industry are unable to exit again without losing a portion of their investment. As a result, hit-and-run strategies are much less profitable and incumbents are able to keep price above average cost. Thus, with sunk costs, markets are not contestable and market power is once again an issue.

## Vertical Structure

In addition to becoming larger through growth and horizontal integration, the major agrochemical companies have restructured themselves in other ways. One of the major changes is increased vertical integration—for example, the combining of seed, chemical and biotechnology activities within the same firm. Although some companies initially tried to include pharmaceutical and biotechnology components within the life science model, practically all of them have now divested their agricultural operations. A further element of this internal restructuring is the increased use of strategic alliances.

Many countries have had their own local seed companies that have over the years developed seed for a specific geographical market and operated sales and distribution systems. The major biotechnology companies are increasingly purchasing these seed companies as a source of seed material in which to insert the genes for herbicide or insect resistance (Kalaitzandonakes & Hayenga, 2000). As an example, in 1997, Monsanto acquired a 30% share of the Brazilian corn seed market with the acquisition of Sementes Agroceres. With its 1998 purchase of Cargill's international seed division, Monsanto now controls over half the Argentine maize seed market. In 1998, Dow AgroSciences acquired Morgan Seeds, Argentina's second largest corn seed company, and Brazil's Dinamilho Carol Productos Agricolas, another key South American corn seed company. Phytogen (majority owned by Dow Agrosciences) acquired a major cottonseed breeding program in the Chaco province of Argentina. In 1998, Mexico-based Empresas La Moderna (ELM) bought two South Korean vegetable seed companies and Nath Sluis (agricultural biotech company) of India (Action Group on Erosion, Technology and Concentration [ETC], 2000).

## Factors That Affect Horizontal Industry Structure

Numerous factors are behind the wave of horizontal mergers and acquisitions described above. Some of these factors are common to all industries and have no specific link to the seed, chemical and agricultural biotechnology industries. Thus, the mergers and acquisitions in the seed and chemical industries are at least in part a result of the need to consolidate costs and rationalize industry capacity, a desire by the management of the firms involved to extend their sphere of influence, and a wish by some firms to pre-empt other firms from taking over valuable assets (Shy, 1996). MacDonald (2001) provides an overview of concentration levels in a number of US agricultural sectors, many of which have seen higher concentration in recent years.

There are at least two explanations for increasing concentration, however, that pertain specifically to the seed, chemical and agricultural biotechnology industries: (a) sunk costs, and (b) escalation strategies.

## Sunk Costs

Two sources of sunk costs appear to be important in determining industry concentration: (a) intellectual property rights and R&D expenditures, and (b) the regulatory requirements that governments have introduced before the products of the R&D activity can be marketed.

On the empirical side, Ollinger and Fernandez-Cornejo (1998) examine sunk costs and regulation in the US pesticide industry. Using data over the 1972-89 period, they find that research costs and pesticide regulation costs negatively affect the number of companies in the industry, and that smaller firms are affected more strongly by these costs than are larger firms. Research and regulation costs also encourage foreign-based firms to expand into the US market and to force less profitable innovative firms to exit the market. Ollinger and Fernandez-Cornejo also point out that their results on the impact of regulatory costs generally match those found in other industries.

These empirical results can be linked to theoretical arguments concerning sunk costs and the economies of scale and scope that sunk costs tend to create. As Sutton (1991) outlines, the presence of sunk costs means that for firms to be profitable, price needs to be raised above marginal cost, typically by reducing the amount of competition (i.e., the number of firms). Sutton identifies two types of sunk costs: exogenous and endogenous. Exogenous sunk costs are those beyond the control of the firms in the industry. Regulatory costs are a good example of exogenous sunk costs. In contrast, endogenous sunk costs are firm-level strategic variables. Advertising and R&D are examples of endogenous sunk costs. The rest of this section examines the connections between intellectual property rights (IPRs), R&D expenditures, regulatory costs, and economies of scale and scope.

Economies of scale and scope are major factors in propelling industries towards concentration. Economies of scale exist when average costs fall as more output is produced; economies of scope exist when the total cost of producing two outputs together is less than the cost of producing the two outputs separately. Because economies of scale and scope mean that larger and diversified firms have lower average costs, there is clearly an incentive for firms to get large (Fulton, 1997; Lesser, 1998; Hayenga, 1998). Indeed, those that do not get large are vulnerable to being driven out of the market by larger and more cost efficient firms. Of course, there is a limit to how large firms can get. Although development and production costs may fall with an increase in the size of the firm, other costs—particularly those associated with administration—rise. Nevertheless, economies of scale and scope clearly create pressures for consolidation.

Economies of scale and scope are created as a result of investment in non-rival goods. Unlike rival goods—such as materials, labour, and energy—which can only be used in one place, by one person and at one time, non-rival goods can be used in more than one place and by more than one person, all at the same time. This feature of non-rival goods—that they can be used over and over again—means that output can be increased without having to increase all inputs. Consequently, economies of scale and scope are created (Romer, 1990; Fulton, 1997). Much of the expenditure on non-rival goods—examples include R&D and regulation expenditures—can be considered a sunk cost.

Intellectual property, of course, is a good example of a non-rival good. Indeed, ideas generally are considered non-rival goods. It is widely believed that many of the high-technology industries, including the biotechnology and information industries, are subject to increasing returns. Romer (1990) postulates that this is a result of the distinction between physical goods and ideas. Ideas are not scarce, and as such, any industry based primarily on the trade of intellectual property will not face diminishing returns in their primary resource, the idea. The protection of intellectual property, through IPRs, is designed to encourage innovation and to provide incentives for the development and diffusion of new products and technologies.

An example illustrates the connection between intellectual property and economies of scale and scope. Suppose a biotechnology firm has some intellectual property, such as a particular gene that has been isolated. This intellectual property can be used over and over again as the firm expands its activities. For instance, if the company wishes to develop seeds for a new crop, it will not have to invest again in the research that isolated the gene. Although the development of a new seed will require additional lab and greenhouse space, labour, and materials, the expenditure on the technological advancements does not have to be made again.

A similar result occurs if the firm needs to invest substantial amounts of money in obtaining regulatory approval for a seed—although the production of additional units of the seed will require additional costs, the regulatory expenditures do not have to be made again. Once again, large companies typically have an advantage, because they are able to spread the costs of obtaining regulatory approval over more output. Thus, the greater the regulatory requirements in an industry, the more concentrated the industry is expected to be.

Intellectual property may also create economies of scope. Once a specific gene has been isolated—for instance, a gene that confers a resistance to a particular herbicide—this gene can then be put in a number of crops. Once again, the production costs of a number of products together will be less than if the products were produced separately.

To recap, R&D expenditures and regulatory costs are both sunk costs and a source of economies of scale and scope. Because economies of scale or scope mean that larger and more diversified firms have lower average costs, there is clearly an incentive for firms to get large. Indeed, those that do not get large are vulnerable to being driven out of the market by larger and more cost efficient firms. These theoretical results are supported by empirical observations in the US pesticide industry.

## Escalation Strategies

An escalation strategy is one in which a company spends large amounts on R&D and engages in mergers and acquisitions to achieve a dominant role in the market—that is, the firm tries to leapfrog its competitors to become the dominant firm. Escalation can be a profitable strategy when there is a high degree of substitutability with competitors' products on the demand side, and there are scope economies on the supply side (Sutton, 1998). Both of these factors are present in the agricultural biotechnology industry.

The isolation of a gene that provides particular advantages and which can be inserted into a number of crops means there are scope economies on the supply side. There are also clear supply side scope economies associated with the enabling technologies that are required to use these genes. And on the demand side, herbicide and insect resistant seeds and the accompanying chemicals are clearly a substitute product for traditional seeds and herbicides and pesticides (Hayenga, 1998).

Firms engage in escalation strategies in a number of ways. Although internal growth through R&D is one way, the more common method is through consolidation via horizontal mergers and acquisitions. As the theory predicts, the combination of demand substitutability and supply side scope economies present in the seed and chemical industry appears to be linked with escalation strategies. Monsanto has clearly followed an escalation strategy, with Dow, Novartis, and others following similar strategies of making acquisitions and spending significant amounts on R&D (Bijman [2001a] notes that agricultural seed companies typically spend 15% of turnover on R&D).

## IPRS AND INVESTMENTS IN PLANT BIOTECHNOLOGY

The implicit rationale for IPR protection is that they promote investments in plant breeding and bio-engineering. This hypothesis is contested by section of analysts. Nevertheless it is a fact that the structure of IPRs do influence investments. IPRs induce their own pattern of innovations. The time taken for R&D to fructify as innovations is a crucial aspect guiding investments.

*Figure: Plant biotechnology.*

Often IPR regimes play a key role in influencing the time pattern of innovations. It is a well-known fact that conventional plant breeding methods are dilatory and time consuming. It requires 7 to 8 generations of repeated breeding to reduce heterozygosity of new genotypes. As Vossen (1985) observes Arabica Coffee subject to three to four cycles of breeding still display heterozygosity and it is not until 20 to 25 years (spanning 4 to 5 generations) that one arrives at true breeding varieties. This time lag is even more pronounced for the Arabusta coffee variety (inter-specific hybrid of Arabica and Robusta Coffee) where segregation of unfavourable characteristics could still occur even after several generations of backcrossing and selfing.

Another interesting example is the one cited by Evenson (1991) about the long gestation period in regard to rice varieties in India. According to Evenson, India released 306 rice varieties for planting during the period 1965-1986 after performing 20,000 crosses over a period of over 15 years since 1950. It must be understood that this long process of backcrossing was preceded by considerable time spent on location and identification of individuals plants with requisite traits. The speed or pace at which plant varieties qualify for being released for planting have economic implications on investment behaviour in seed/plant breeding industry. Asexual modes of plant breeding which include plant genetic 'engineering' and transgenic techniques could obviate the long gestation period associated with pre-release preparations encountered conventional breeding.

*Figure: Through an innovative feat of plant biotechnology and vaccine design.*

It is a different matter than biosafety concerns may delay releases of these varieties. The time lag in release of plant varieties fulfilling the criteria of novelty, distinctness, uniformity and stability (or alternatively the requirements of novelty, inventiveness and commercial application) raise fundamental issues from the view-point of capital investments. The time lag in varietal release is potentially least in the case of transgenic and non-sexually produced varieties while for traditional or classical methods the time lag can be greater.

The UPOV 1991 based plant IPR regimes could induce longer time lags as compared to the situation when an IPR regime is based on the 1978 version of the UPOV. This is due to the fact that the minimum genetic distance concept implicit in UPOV 1991 precludes 'close cousin' varieties from acquiring protection rights, unless the techniques of genetic distance determination by themselves are inadequate. UPOV 1978 would facilitate capital investments to some extent as the release of plant varieties is periodically more frequent. Even under the plant patent regimes, investments could bring about short-term returns, though the volume of investments will have to be larger on account of capital intensity of advanced plant genetic engineering technologies. By comparison, UPOV 1991 regime is least conducive to capital investments as plant variety releases can be painfully slow under this regime. Therefore sustenance of conventional plant breeding under UPOV 1991 regimes, can only be ensured by infusion of low discount capital.

Paradoxically it is the reduced availability of low discount public funds for plant breeding that has induced countries like India to go in for Plant Varieties Protection Legislations modelled on UPOV lines. The implicit assumption in undertaking this step is that private funds for R&D in plant breeding will 'crowd in', if protection plant varieties rights are accorded. But the reality is that in liberalised deregulated developing economies, private investments have 'crowded in', in short return promising avenues.

Consequently classical plant breeding activities will suffer for want of funds in the UPOV 1991. In other words, UPOV 1991 modelled national legislations will not necessarily hold attraction for private sector investors in R&D projects involving classical plant breeding as compared to say, UPOV 1978 or the Plant Patent regimes. Rather private investments would be catalysed better by *sui-generis* plant protection legislations, which partake of plant patent elements and focus in the main on hybrids and genetically engineered varieties.

Given the linkages between plant breeding and the seed industry these perverse investment behaviour trends can produce major economic consequences. The Seed Laws in various countries, both advanced and developing, are getting to be progressively oriented towards restricting market circulation to varieties that are conferred with plant breeder or patent rights. The UPOV guidelines on DUS are used world wide not only as the basis for establishing varietal distinctiveness and descriptions but also for seed certification purposes. The DUS guidelines are employed for recognising and registering not only the 'basic' seeds used for multiplication of 'pedigree' seeds but also for certification of the pedigree seeds themselves.

## IPR REGIMES FOR PLANT BIOTECHNOLOGY

Article 27(3) of the TRIPs Agreement lays down that members shall provide for the protection of plant varieties either by patents or by an effective *sui-generis* system or by any combination thereof. Article 27(3) in particular is critical to India as this requires us to provide protection of plant varieties by conferring plant breeders rights either by way of patents or through *sui-generis* plant variety protection legislations. Countries with strong R&D base in plant genetic engineering such as USA have robust Plant Utility Patent Legislations. India is certainly not inclined to adopt patent protection regimes for its plant varieties. India is rather inclined to adopt a *sui-generis* legislation which is non-'patent' based. There are many reasons for this proclivity. India is one of the ten mega-diversity countries of the world and a rich storehouse of landraces of principal agricultural crops. India has a strong R&D base in conventional plant breeding methods. Its strength in plant genetic engineering is impressive but not an overwhelming factor by comparison. The first two strengths explain India's disinclination towards plant patent regimes or towards a *sui-generis* legislation which is 'patent'-driven.

While Plant Utility Patents Act provide for broad patents over plant varieties, traits and genes and even the physical parts of the plants, plant breeders rights provide for IPR only over varieties. As is well known, since 1990s the UPOV (Union for the Protection of Varieties) has been largely viewed by developing countries as offering the best regime for positioning their national legislations.

The central feature of the UPOV is the protection it affords to plant breeders who produce plant varieties that fulfil the criteria of distinctiveness uniformity and stability (DUS). The current version of the UPOV, viz. UPOV 1991 had added additional criteria of 'new' to DUS thus rendering DUS as NDUS. Contrary to the popular notion, the NDUS criteria of UPOV 1991 is not substantively different from the principles of 'novelty', 'inventiveness' and 'industrial application' (NII) which applies for patents. The criteria of 'novelty' and 'inventiveness' in Patent Laws are covered by the criteria of 'new' and 'distinct' in UPOV 1991. Thus by distinctness, the UPOV means a variety of plant which is 'clearly distinguishable from other varieties whose existence is a matter of common knowledge'. It is apparent that this term captures the attributes of 'novelty' and 'inventiveness' implicit in Patent Laws. Even in respect of 'uniformity' and 'stability' criteria the UPOV does not offer different recipes. True, by 'stability', the UPOV conveys that 'relevant characteristics of protected plant variety remain unchanged either for a specified period or after repeated propagations or cycles of propagations'. It is also true that 'stability' is a difficult criteria for a plant breeder to fulfil. Attainment of 'stability' criterion is problematic for cross-pollinated plants and non-single homozygous lines of autogamous plant varieties. This, in turn, reduces the commercialisation potential of the plant variety since inconsistency of genetic quality jeopardises commercial application of the plant variety. The same holds true of the criterion of "uniformity". Therefore, the NDUS criteria of UPOV 1991 are homologous to the NII criteria implicit in Plant Utility Patent Laws.

The UPOV has undergone two major amendments since its inception in 1961, i.e. in 1978 and 1991. The main concern with regard to currently operational UPOV 1991 has been its stringent provisions, which include the additional attribute of 'new' to the DUS criteria. The scope of UPOV 1991 in terms of coverage is larger. While plant varieties of all taxa are covered by UPOV 1991 only plant varieties of nationally designated species were covered by UPOV 1961 and 1978. Further by defining clearly that 'variety' represents a 'group of plants', or 'single' or 'several plants' or 'one' or 'several parts' of a plant(s) the UPOV 1991 has considerably enlarged the scope of its coverage. Going by this definition even sexual and vegetatively propagated varieties are covered by UPOV 1991. The US PVPA legislation has been quick to adopt the UPOV 1991 formula. The amendments of 1994 to the US, PVPA broadens the term 'plant varieties' to include all materials harvested from the protected varieties. The other drastic feature of UPOV 1991 has been abridgement of the 'plant breeders exemptions' and 'farmers privileges' conferred by the earlier versions.

Earlier versions of UPOV allowed farmers to re-use seeds generated from their farms for self-use, and allowed breeders to freely use protected varieties for further improvements. These concessions gave UPOV a 'less stringent' face in comparison to 'patent' laws. UPOV 1991 gives an option to National Governments to disallow the farmers privileges of retaining or re-using seeds for self cultivation. UPOV 1991 also restricts the breeders rights by enlarging the right of the breeder to close variations of his cultivar or 'essentially derived varieties'. In other words, the breeder of a cosmetically bred variety would have to buy genetic dependency rights from the derived variety prior to commercialising the derived version. Further the duration of protection have also been extended under UPOV 1991 to 20 years for crops and to 25 years for trees and vines. Thus in terms of these rigorous features, UPOV 1991 eliminates the arbitrage between plant varieties protection and patent regimes.

## NEW TRENDS IN AGRICULTURE

Since time began we humans have always relied on plants and animals to provide us with food, shelter, clothing and fuel, and for thousands of years farmers have strived to better these in order to continue providing us and to meet the needs of evolution. As the population grows and expands then so does the need for resources provided by plants and animals.

With the population expected to reach over 10 billion by the year 2030 it is estimated that the world food production will have to double on the farmland that exists today, if it is to compete with the anticipated growth in the population. The ever-increasing needs of yields of crops and the decreasing crop inputs such as water and fertiliser can all be helped by biotechnology as can advancements in providing pest control that is better for our environment.

## POST R&D REGULATIONS AND INVESTMENTS

Post-R&D regulations play a critical role in deciding on the pace of investments in the biotechnology sector. Even if a start-up enterprise enjoys a favourable IPR regime, that rewards its inventions, it is unlikely that it will invest in product development, in case post-R&D regulations are rigid and dilatory in nature. This then forms the backdrop for strategic alliances between start-up ventures and downstream industrial complexes. Multi-layered regulations in the post-R&D phase involve complexities which are expressed in the delays in commercial application /utilization of invented products or technologies.

In the case of third generation plant biotechnologies, such as transgenics, the problem is compounded by the fact that regulations by themselves are complex and less understood by the regulators themselves.

A transgenic plant would have to undergo multi-stage regulatory checks and clearances before finding commercial application.

India is uniquely situated as far as biotechnology investments and infrastructure are concerned. In the area of plant biotechnology, advances have been made in transgenics of rice and wheat carrying stress tolerant genes such as CodA, COR47 and HVA1. The larger challenge is to take these products through the regulatory processes. More formidable is the issue of market development and appropriate pricing of transgenic seeds. There are many imponderables here. Market shares for seeds in India have varied from crop to crop and from region to region. The 'demand' for transgenic plants or seeds may be even more uncertain, given the 'safety' dimensions and risk aversion propensity of Indian farmers. In the dry agricultural pockets of Central and South India, farmers tend to be attached to traditional drought-hardy varieties of plants, which in turn contributes to their initial 'inertia' to shift to transgenics.

Thus the task of marketing and pricing of novel biotechnology plants gets to be complex in such environment of 'demand incoherence'. A high margin pricing strategy for seeds may choke demand for transgenics. Thus the degree of risks in commercializing transgenic plants could be of a high order. Given these realities, the idea of strategic alliances between start-ups and downstream entities looks to be realistic. The fact that premier R&D institutions such as the Indian Institute of Science, has set up its own start-up biotechnology venture, renders a strategy of tie-ups, a logical option. Much will depend upon the complexities inherent in product development and the willingness of competent downstream entities to signup agreements. To a large extent, this will depend upon enabling policies at the upper end of the value chain.

To sum up, India's agricultural biotechnology industry has the necessary policy and regulatory support to promote active R&D. However, in the critical area of product development and commercialization much more attention needs to be paid. The multi-tiered regulatory framework that exists could seriously affect the performance of start-up companies that would like to run through the entire life-cycle of biotechnology product development.

Consequently, partnerships and strategic alliances should be encouraged between start-up ventures and established companies that have high product development capabilities. The experience of other countries in the Asia Pacific region could be a key input in framing development strategies for biotechnology products in India.

## AGRICULTURAL BIOTECHNOLOGY

Every year, millions of people worldwide die from starvation and nutritional deficiencies. Al-though these challenges remain significant, we are making some progress. Agricultural technologies have increased caloric intake on a per-capita basis in nearly every nation of the world. technological development— particularly in the area of agricultural biotechnology — is crucial if we want to further reduce malnutrition and starvation and meet the needs of growing populations. Unfortunately, misperceptions about biotechnology have led to activist movements to slow down, if not eliminate, this critical technology via government regulation. In the United States, the advancement of biotech crops has slowed down because of attempts to regulate them differently than conventionally grown foods. In Europe, antibiotech movements are much stronger, and international bureaucrats are pushing various international agreements and trade restrictions. While Western nations may be able to absorb the costs of such policies, people in the developing world will pay a higher price as many continue to suffer from starvation.

## Understanding, characterizing and managing genetic resources

Farmers and pastoralists have manipulated the genetic make-up of plants and animals since agriculture began more than 10 000 years ago. Farmers managed the process of domestication over millennia, through many cycles of selection of the best adapted individuals. This exploitation of the natural variation in biological organisms has given us the crops, plantation trees, farm animals and farmed fish of today, which often differ radically from their early ancestors.

The aim of modern breeders is the same as that of early farmers-to produce superior crops or animals. Conventional breeding, relying on the application of classic genetic principles based on the phenotype or physical characteristics of the organism concerned, has been very successful in introducing desirable traits into crop cultivars or livestock breeds from domesticated or wild relatives or mutants. In a conventional cross, whereby each parent donates half the genetic make-up of the progeny, undesirable traits may be passed on along with the desirable ones, and these undesirable traits may then have to be eliminated through successive generations of breeding.

With each generation, the progeny must be tested for its growth characteristics as well as its nutritional and processing traits. Many generations may be required before the desired combination of traits is found, and time lags may be very long, especially for perennial crops such as trees and some species of livestock. Such phenotype-based selection is thus a slow, demanding process and is expensive in terms of both time and money. Biotechnology can make the application of conventional breeding methods more efficient.

# 5

# Livestock Production and Crop Rotation in Green Manures

## LIVESTOCK ON ORGANIC FARMS

Among the thorniest of issues swirling around the edges of organic agriculture is the role of livestock. The disagreements arise because of the diversity of people and philosophies in the organic community. Organic agriculture can usually count vegetarians and animal welfare proponents among its more vocal supporters. Many of these people feel strongly that animals should not be exploited. Their rationale often goes beyond emotional and religious beliefs; convincing human health concerns, social issues, and environmental reasons are commonly cited. On the other side of this argument are those who feel that an organic farm cannot achieve its full potential or ecological balance without livestock manure; that it is essential to nutrient cycling and to the finer aspects of soil building.

Excellent soil fertility can be built in the absence of farm livestock and livestock manures by using vegetation-based composts and by harnessing the livestock in the soil — earthworms and other soil organisms. However, it is clearly easier to design a contemporary, low-input organic farm when traditional livestock are integrated. The biological and enterprise diversity that livestock can bring contributes enormously to stability and sustainability. A good example is provided by Rivendell Gardens in Arkansas, which began integrating livestock enterprises after several years as a solely horticultural operation.

The owners of Rivendell, Gordon and Susan Watkins, now rotate their strawberry and vegetable crops with grass-fed, direct-marketed beef and pastured poultry. Ideally, poultry follows beef on pasture to reduce cattle parasites. The seasons in mixed legume/grass pasture leave the soil quite mellow and well-manured for subsequent high-dollar horticultural crops.

The Rivendell operation demonstrates the sort of organic management where a large number of organic farmers and many animal welfare proponents find common ground.

The Watkins' animals are all raised with minimal confinement and generous access to sunshine, fresh air, and free-choice foodstuffs. While domesticated and destined for slaughter, they lead low-stress lives in conditions much closer to natural than the conventional alternatives. This is the antithesis of industrialized factory farming systems, which are increasingly becoming the norm in livestock production.

Many in both the organic and animal welfare communities are working to prohibit factory farming of livestock in organic systems. Many of the difficulties revolve around the fine interpretations of language in various organic standards. Wording such as "access to fresh air and sunlight," for example, can be construed to mean nothing more than opening the door on one end of a large confinement poultry house for a couple hours a day.

What Howard had not taken into account is the almost universal contamination of urban wastes with heavy metals and chemicals that are not eliminated by composting and may even be concentrated. Perhaps this was not yet a serious problem in his time; it is, however, in ours. Organic farmers and consumers concerned about contamination of soil and crops with agricultural pesticides and synthetic fertilizers would be remiss to ignore the contamination hazards of even well-composted sewage.

Fred Kirschenmann, a farmer and former NOSB member, has written eloquently about the progress of the National Organic Programme. In a critique of the March 2000 draft of the proposed rule, he pointed to another reason why the use of biosolids ought to be prohibited in organic production. Because of the manner in which biosolids are generated,

they are easily hauled and land-applied on an industrial-scale to industrial scale organic farms. Furthermore, since biosolids can essentially supplant animal manures as a source of organic matter and nutrients, their use would allow some very large farms to circumvent the traditional practices that promote biodiversity and enterprise diversity and integration. What Kirschenmann fears from biosolid use is technology that would nudge organic agriculture down the same road of industrialization taken by conventional ag.

### *Intercropping and Companion Planting*

Interplanting two or more mutually beneficial crops in close proximity is one strategy for increasing biodiversity. In large-scale mechanized crop culture, this is called *intercropping*. It typically involves alternating rows or a number of rows of compatible field crops, like soybeans and corn. It also applies to sowing multiple forage crops, like alfalfa, bromegrass, and timothy, when these are grown together.

When interplanting is done on a smaller scale, it is often called *companion planting*. A classic example of companion planting is the inter-planting of corn with pole beans and vining squash or pumpkins. In this system, the beans provide nitrogen; the corn provides support for the beans and a "screen" against squash vine borer; the vining squash provides a weed suppressive canopy and discourages corn-hungry raccoons.

### *Biological Pest Control*

Organic farming relies heavily on populations of beneficial insect predators and parasites, pest disease agents, insect-eating birds and bats, and other creatures, to help manage pest problems. These biological controls help keep pest numbers at levels where further cultural activities or relatively mild pesticides are (usually) adequate to assure a crop. In some instances, biological control can be so effective that no additional action is even needed by the farmer.

Some see biological control as a default benefit of the soil fertility practices of organic farming. The diversity of crops in a soil-building rotation, the use of cover crops, and other practices build a diverse soil biology that works to keep soil pests in check. They also provide substantial above-ground habitat for beneficials. The absence of pesticides also favours biocontrol. In many organic systems, farmers sometimes purchase and release control agents like ladybird beetles, lacewings, trichogramma wasps, etc., or use weeder geese — a quaint but effective biological weed control. Increasingly, growers are designing and maintaining both permanent and temporary habitats specifically for beneficial insects, spiders, and other helpful species. This is known as *farmscaping*.

### *Sanitation*

Sanitation can take on many forms:

- removal, burning, or deep plowing of crop residues that could carry plant disease or insect pest agents

- destruction of nearby weedy habitats that shelter pests
- cleaning accumulated weed seeds from farm equipment before entering a new, "clean" field
- sterilizing pruning tools.

As in human and animal health, sanitation practices can go a long way in preventing crop pest problems. However, many practices — such as clean cultivation, deep plowing, and burning crop residues — can increase erosion and reduce biodiversity. Thus, they may conflict with sustainability. Good organic growers recognize this and treat those practices as transitional or rescue options, rather than relying on them on an annual basis.

### *Tillage and Cultivation*

Tillage and cultivation are tools that can accomplish a variety of objectives in farming systems: weed control, crop residue management, soil aeration, conservation of manures and other fertilizers, hardpan reduction, sanitation to destroy pest and disease habitat, etc.

While conventional farmers rely on chemicals to accomplish many of these objectives, organic growers focus more on improving tillage and maximizing its benefits. Guidelines for primary tillage, for example, are intent on conserving crop residues and added manures in the upper, biologically active zones of the soil, rather than burying them deeply where decomposition is anaerobic (oxygen-starved). Leaving soils completely bare and vulnerable to erosion is discouraged; fall moldboard plowing is certainly frowned upon.

Cultivation in organic systems often rises to the level of art. Row-crop farmers frequently use *blind cultivation* —shallow tillage, which largely ignores the crop rows—beginning shortly after seeding until the plants are but a few inches high. Rotary hoes, wire-tooth harrows, and similar equipment can be used for blind cultivation, delaying the first flush of weeds and giving the crop a head start.

## INTENSIVE LIVESTOCK PRODUCTION

The intensification of livestock production has had not only a significant effect on animal welfare, but also a considerable impact on the environment. In this regard, the disposal of the enormous quantities of animal waste produced by animals confined in intensive livestock production units has created the greatest environmental problems. The reduction of manpower levels and the need for efficient animal management have meant that 'factory' farmers have often abandoned the use of straw as bedding in livestock houses; the animals instead stand on slatted or solid concrete floors. The faeces and urine which these animals produce must therefore be disposed of differently.

When straw bedding is used, manure is collected in dung heaps, exposed to air it gradually ferments, bacteria are thus to some extent destroyed and the excreta eventually becomes fertilizer to be spread on the land. Slurry, on the other hand, must be stored in tanks, where it may decompose and sometimes produce a variety of noxious smelling gases. This method does not kill pathogens quite as effectively as the traditional method.

While straw bedding most closely resembles the natural surface on which most animals lie—and is thus promoted as the ideal floor surface on which animals should be kept—it can provide the conditions under which disease organisms can flourish. If poorly maintained, such bedding when excessively dirty and wet not only causes distress to the animals, but can also lead to environmental mastitis in dairy cattle, 'farrowing fever' in pregnant sows and an increased exposure of poultry to the dangerous *E.coli* bacteria.

Slurry can be returned to the land as fertilizer, but its storage and application can be particularly problematic. It cannot be spread when the ground is too wet and must, therefore, be stored in huge tanks, sometimes for several months. The failure of these storage tanks would have a significant impact on the local environment, most particularly if pathogens, such as salmonella, *E.coli* and leptospira, which are dangerous to both human and animal health, were still present in the slurry. To reduce the risk of infection, farmers have been warned against spreading slurry on grazing land.

If they must do so, they are advised to allow the slurry to adequately ferment and to leave the land on which it has been spread unpastured for at least six weeks afterwards. The main problem with modern farms is that they do not always have enough land to absorb the waste that the animals produce. The waste must, therefore, often be transported elsewhere at considerable expense to the livestock holder. As Johnson suggests, waste disposal can be the main factor which determines how many animals can viably be kept in a livestock production unit, rather than animal welfare. To illustrate this problem, he describes the crisis faced in the Netherlands during the late 1980s.

The great concentration of factory farms had put so much pressure on land that by January 1987 the Dutch authorities placed a ban on the further expansion of livestock numbers. With farms producing around 100 million tonnes of effluent per annum, twice the amount of liquid waste accumulated than could safely be spread on the country's 2 million hectares of farmland.

As a consequence, the transport of slurry to areas away from factory farms was encouraged and subsidised by the Dutch government and large-scale dumping of slurry was banned due to the potential pollution and health risks it presented. Dutch farmers are now required to pay the full costs of disposal through sewage plants if they are unable to find appropriate land to spread their farm waste on.

Like slurry, the production of silage—a grass crop harvested when green and then partially fermented to make animal fodder—can present serious environmental problems. The anaerobic fermentation which accompanies silage production can potentially produce a waste liquid which has been estimated to be 200 times more polluting than domestic sewage, particularly when the grass is wet. Such waste presents great pollution problems if it escapes the silos in which it is held and enters natural waterways. The aquatic life of rivers and streams can be either poisoned or starved of oxygen as a result of the introduction of this effluent.

The contamination of natural waterways is one of the most serious environmental consequences of modern agricultural practice. Aside from the problems posed by slurry and silage effluent, intensive livestock production contributes towards nitrate pollution. In addition to this, the additives used in animal feed and the antibiotics which the animals are routinely given are potential sources of environmental contamination. Although the blame for environmental pollution is often apportioned to industry, rather than agriculture, the intensification of farming methods—most particularly with respect to the waste it produces—presents a significant threat to the environment.

In addition to the phenomenal impact on animal welfare and the environment, the intensification of livestock production has created new opportunities for disease organisms to flourish and threaten the health of both animal and human populations. As this book has thus far clearly illustrated, our interactions with other animals have always brought with them the potential risk of infectious disease. However, the increasing exploitation of and dependency on animal resources, which has occurred through the intensification of livestock production, has meant that disease *risks* which did not exist, or were unthinkable, forty to fifty years ago are today commonplace. In short, the increase in stocking densities has led to the increase of opportunities for disease transmission.

The housing and living conditions of intensively farmed animals—most notably pigs and poultry—potentially provide significant exposure to bacteria and viruses, thereby enhancing the risk of disease. Disease risks can be greatly reduced with good stockmanship. On 'factory farms', however, animal management is primarily linked to maintaining or increasing productivity. In the past, stockmen could attend to animals more individually and were familiar with the small herds they tended.

Today, stockmen are not only fewer in number, but are also required to manage huge numbers of livestock as efficiently and cost effectively as possible. The reduction of manpower levels in livestock farming combined with the increasing numbers of animals kept on site can lead to inefficiency in disease control and hygiene which, in turn, may result in the outbreak of infectious disease amongst the animal stock and—depending on the nature of the disease in question—possibly even transmission to the humans who come into direct contact with them. This is, of course, not to mention the risk of contaminated animal protein entering into the food chain and putting human consumers at risk of infection.

One of the disease management techniques adopted in intensive livestock production is to accept the pervasiveness of infection. Antibiotic drugs and vaccinations are thus routinely administered in order to keep stock healthy and maintain productivity. When disease outbreaks occur, it may not be viable to cull stock given the size of the animal population housed. Similarly, farm buildings may have been too poorly constructed and cannot be properly disinfected or cleaned, the animal's excreta, as suggested above, creating potential hazards for both animal and public health.

Management of disease purely through drug intervention is far from ideal. Moreover, it arouses considerable consumer concern about the quality of the animal produce which ends up on our dinner plates. The use of growth hormones and the possibly harmful effects of veterinary drug residues in food have become an increasing source of consumer anxiety in recent years; as has the suggestion that the persistent administration of antibiotic drugs may ultimately lead to the evolution of pathogens resistant to such substances. For fifty years, antibiotics have been added to animal feeds in order to improve feeding efficiency and growth rates. Since the 1960s, scientists have looked into the possibilities of resistance risks and by 1970 the European community had made moves to prevent the antibiotics that are used in human medicine being used as animal growth promoters. Today, the issue of resistant risks has again become the topic of political and scientific debate, stimulated in part by the development of a veterinary antibiotic called *avoparcin* which is closely related to an important human medical antibiotic known as *vancomycin*. There is a concern that such veterinary antibiotics may promote the emergence of bacterial strains that have a genetic resistance to antibiotics.

Although there is currently little microbiological consensus on this topic, some scientists have gone so far as to suggest that the widespread use of antibiotics in livestock production, particularly those used to treat infections, may well be responsible for resistant strains of salmonella and campylobacter that are being found in increasing numbers of food poisoning cases in humans. In recent years, the problem of animal disease control has also been tackled from another angle, that rather than seeking to alter the living conditions of the animals, instead seeks to alter the animals themselves. In this context, attempts have been made to rear strains of 'minimal-disease' animals which are kept in highly controlled environments. Poultry flocks today can be bred to be 'specific-pathogen free', which essentially means that particular disease-producing agents have been eliminated by breeding and rearing birds in sterile conditions. Such isolation, however, can render such birds vulnerable to other microorganisms that might otherwise have been innocuous.

Modern intensive livestock production is a risky business and sometimes it is human health that is at stake. In spite of the industry's endeavours to produce animal products that are safe for human consumption, as indicated above, food-borne infections still pose a threat to public health. The intensification of poultry rearing is the main culprit for bacterial infections in humans. Salmonella infection is one of the ever-present hazards of broiler rearing. It is in fact widely accepted that salmonella will persist in poultry stocks, although it can be kept at bay by good feeding and housing practices.

The pervasiveness of this bacteria in poultry is largely due to contaminated feed, although people and wild animals or birds can also be the vectors of transmission. Intensively reared poultry, like other farm animals which can carry the salmonella bacteria, are generally fed on high-protein feeds which can contain slaughterhouse waste and other animal by-products such as dried poultry manure. If such feed is inadequately sterilised, disease organisms can persist and can be passed on to other animals. Sterilisation is a costly process which occurs through heat treatment. However, irradiation of animal feed is not always efficient because strains of salmonella have emerged which are heat-resistant and continue to contaminate feed.

The transmission of salmonella requires 'the transfer of faeces or intestinal contents from an infected individual to the digestive tract of a susceptible one'. Like animals, humans can become infected by salmonellosis through the ingestion of contaminated animal produce. The disease can cause gastroenteritis, septic shock syndrome and focal infection in humans, and can even be fatal to vulnerable individuals, such as infants or elderly people. It should be emphasised that good hygiene and effective cooking methods from the abattoir to the kitchen can reduce the risk of infection for the human consumer. The modern systems of livestock rearing, mass animal transportation and conveyor belt style slaughterhouses, however, increase the possibilities of cross-infection and the contamination of human food. Although the most common bacterial infection, salmonella is not the only disease which may be contracted from the ingestion of animal produce. Bovine tuberculosis, brucellosis, trichinosis, campylobacteriosis, listeriosis, staphylococcal enterotoxicosis, clostridial toxicosis and toxoplasmosis may be transmitted through tainted animal food and cause human illness.

Aside from food-borne infections and intoxications, consumer health is generally not affected by the diseases which farm animals carry. The people at greatest risk from agricultural animals and their products are in fact those who work with them; in other words, farm workers, veterinarians and individuals who handle animal carcasses occupationally. In Western Europe today, the major zoonoses, such as brucellosis (contagious abortion), tuberculosis and anthrax, are uncommon.

Other zoonotic diseases, however, present serious occupational hazards to those in the meat, dairy and animal product industry. For example, leptospirosis—commonly known as Weil's disease—is a disease which can seriously affect dairy farmers, both economically and physically. It is a contagious bacterial disease which can produce a high abortion rate in cattle late in pregnancy; it can be responsible for high mortality in calves, reduced milk production and contaminated milk. The organisms that cause leptospirosis are transmitted through urine and live in water or moist soil. Cattle—and other livestock—become infected through being splashed with infected urine, grazing on contaminated pasture or by drinking infected water.

Other wild species, particularly rodents, are responsible for transmitting the disease to cattle, and humans can be infected by bathing in contaminated water or direct contact with infected urine. The effect of leptospirosis on humans varies from inapparent to acute febrile illness which can be accompanied by fever, hepatitis, jaundice and meningitis—amongst other maladies. Good animal management, vaccination and reduction of contacts between livestock and rodents reduce the risks of infection by leptospirosis. There are many more health hazards for those who come into routine contact with food-producing animals. Other occupational hazards for humans are presented by rickettsial diseases, fungal infections (*e.g.* ringworm), chlamydia infections and respiratory viral diseases, in addition to bacterial infections such as those described above.

The fact that intensive livestock production today happens within closed systems, away from public view, has much to do with the reduction of infection risks. Modern livestock housing is essentially designed to keep out disease-causing organisms and their vectors of transmission and to keep disease-free animals in. Such systems are far from foolproof; infection may, for example, occur through the drinking water or animal feed. Furthermore, farmers, stockmen and veterinarians may introduce disease to livestock. It is commonplace when one enters a livestock production unit for one to have to don special overalls and boots in order to minimise the possibility of introducing new pathogens to the animals from outside. However, keeping all visitors out, apart from stock and veterinary personnel, not only aids disease prevention and control, but to some extent prevents consumers from gaining a first-hand glimpse of the conditions under which their food is produced.

Modern consumer society demands a continual, cheap and dependable supply of animal products which are pre-packaged and ready to enjoy: yet never before have the majority been so divorced from the production of the animal proteins which sustain them. The urban consumer is estranged from the origins of the food he eats and the leather he wears and has limited involvement in or knowledge of the circumstances of their production. The consumer is very much reliant on the producers and retailers to offer fresh and uncontaminated animal products—at a low price—on demand. The 'farm fresh' animal products which are nowadays found in urban supermarkets are however unlikely to have been produced locally, as was formerly the case. Animal produce may originate from farms and factories situated hundreds of kilometres from the consumer's domicile. The transport networks and storage capacities which today so efficiently ensure the swift despatch and delivery of farm-fresh animal produce to the consumer were not possible until recently.

Previously the movement of livestock and carcasses was fairly restricted, both within and across provincial or national borders, not only through lack of adequate transportation and refrigeration, but also due to the fear and possibility of the spread of contagious disease. Once a sufficient level of understanding and knowledge of transmission and control through governmental intervention had been reached, disease could be better contained, cross-infection prevented and animals could be more safely moved from one location to another.

Yet although governmental regulations pertaining to animal disease control have existed both nationally and internationally since last century, it is still possible for contaminated animals to slip the net and infect others (be they animal or human). Given that infected livestock or tissue can easily and quickly be moved from one (international) location to another, new opportunities for disease-producing organisms can develop in areas and amongst animal-human groups previously unaffected. The risk of disease can be further exacerbated by the failure to properly disinfect vehicles used for animal transportation, and delays in the transfer of live animals or meat from one place to another facilitate contagion further. Since, on an international level, animal movements are more likely to increase than decrease as a result of political and economic accord, the potential problem of animal disease transmission could, in the future, be one of increasing significance.

## Veterinary Regime during the Twentieth Century

As the above discussion suggests, the possibilities for the transmission of infectious diseases are very real, most particularly given the risks posed by intensive systems of agricultural animal management and animal transportation. It is largely for this reason that veterinary medicine has become so important during the twentieth century in maintaining and monitoring animal health, to keep the spread of infectious disease in check and to ensure that farmers adhere to and understand the guidelines and regulations concerning disease prevention and control. Yet while disease *risks* may today be far greater than ever before, the danger of humans *actually* contracting diseases from contact with animals or the consumption of animal produce is in fact relatively small.

Throughout the course of the twentieth century, the veterinary regime has continued to move from strength to strength. The state veterinary services that were instituted in the industrialised nations of Europe during the late nineteenth century have played an increasingly important role in the control and prevention of animal disease. Legislative measures to combat the effects, spread and introduction of contagious disease have been enforced with increasing vigour. The successes of earlier campaigns and stringent legal controls to stamp out the animal plagues that devastated European livestock populations provided the basis for future legislation. For example, veterinary inspection and the rigourous enforcement of all regulations pertaining to the importation, transport, sale, slaughter and destruction of livestock successfully led to the eradication of contagious bovine pleuropneumonia in the Netherlands by 1887 and Britain in 1898.

The introduction of measures to financially compensate farmers for their losses through compulsory slaughter policies undoubtedly provided an incentive to abide by governmental regulations and allowed veterinary officials to perform their duties without hindrance. Eventual success in eradicating disease helped to persuade livestock producers to accept the ruthless and restrictive regulations that the state sought to enforce. As a consequence, state veterinary intervention has become institutionalised within modern livestock production. Moreover, during the twentieth century, the appreciation of the need for animal disease control and prevention has more or less been incorporated into the habitus of livestock producers.

Scientific developments have, of course, played a crucial role in the prevention and control of animal disease. The development of the science of bacteriology, advances in pathology, parasitology, immunology and microbiology and so forth have provided important contributions to advances in animal and public health. By the turn of the century, European governments had become increasingly aware of the need for better scientific understanding of animal disease if they were to reduce the risks that it posed. In the Netherlands, for example, the State Serum Institute was founded in 1904-by the pioneering veterinarian and bacteriologist Jan Poels—with a view to steering the work of the established Dutch Veterinary Service towards more scientific means of preventing and curing livestock disease.

The Serum Institute experimented with the production of a serum against foot-and-mouth disease and an anthrax vaccine. Moreover, it conducted bacteriological and chemical investigations into milk and dairy produce. Interestingly, this new institute operated specifically outside the dominion of the veterinary school, being instead an organ of the state. In 1933, nearly three decades later, the State Veterinary Research Institute was established, primarily to study and develop a vaccine against foot-and-mouth disease—a highly infectious disease that has perhaps wreaked the greatest havoc on European livestock and the livestock industry during the twentieth century. By 1946, the Institute's director, H.S. Frenkel, had succeeded in developing a method by which a vaccine against foot-and-mouth disease could be produced in large quantities. This method, however, only proved effective until 1951, when a more virulent strain of the disease struck Dutch livestock.

From 1945 onwards, the structure of the state veterinary machine became increasingly more complex, as government-subsidised Animal Health Services were established in each province within the Netherlands. The earliest regional animal health service had in fact already been established in Friesland in 1919. Its early work led to significant progress in the battle against bovine tuberculosis. The new regional services were modelled on their Friesian predecessor and strove for the coordinated control of animal disease, involving not only scientific and governmental veterinary advisors, but also representatives of local farmers and dairy organisations and practising veterinarians.

The eleven regional Animal Health Services were in turn coordinated by the Industrial Board for Agriculture, who established an Animal Health Committee to deal with the allocation of funds and organisation of schemes for disease control. After the Second World War, the regeneration of the meat and dairy export industry became an important part of Dutch economic recovery. A key aspect of this was a national plan to rid Dutch cattle herds of tuberculosis, thereby ensuring the safety and quality of the national product. A considerable portion of the Marshall Plan funds that were made available to the Netherlands to assist post-war economic reconstruction was devoted to achieve this end. By 1956, the disease had been conquered due to the efforts and cooperation of the Animal Health Services, practising veterinarians and the meat inspection service.

On the basis of this national plan's success, similarly successful attempts were made to control and eradicate brucellosis from the Dutch livestock population. By 1957, the Serum Institute and the State Veterinary Institute amalgamated to form the Central Veterinary Institute, which functioned under the auspices of the Dutch Ministry of Agriculture and Fisheries. The Veterinary Service was effectively given the governmental mandate, through the establishment of legislation such as the Livestock Act and the Veterinary Medicines Act, to prevent animal disease and preserve public health.

The work of the Central Veterinary Institute continued to grow and diversify throughout the 1960s and 1970s. In 1986, it consolidated and moved to new premises in Lelystad. By 1994, the CVI amalgamated with a number of other agricultural animal research institutes to create the Institute for Animal Science and Health. Today, this institute employs over 600 people and conducts research into animal health, breeding and reproduction, animal welfare, stress and management, animal nutrition, and so forth. Moreover, it is responsible for diagnostic work, the production of vaccines, specific-pathogen free animals, diagnostic preparations, etc., and the control and standardisation of veterinary medicines.

The developments in state-directed animal disease control in the Netherlands can be mirrored in the rise of the State Veterinary Service in Great Britain. There too was an understanding for the necessity of improved scientific knowledge and diagnosis of animal disease. The first governmental diagnostic laboratory was set up in 1894 in response to an outbreak of swine fever. This was followed by the establishment of the Central Veterinary Laboratory at

Weybridge, Surrey, in 1913. Initial research was conducted into spontaneous abortion, swine fever, scrapie, foot-and-mouth disease and diseases of poultry.

As farmers began to increase the size of their flocks, as the process of intensification of production began to take off, the laboratory found itself increasingly dealing with the effects and greater incidence of poultry disease. During the 1930s, the laboratory began to research and produce tuberculin, in addition to an anti-abortion vaccine. The Agricultural Act of 1937 facilitated further expansion of the laboratory. New research departments of bacteriology, biochemistry, parasitology, pathology and poultry disease were established and a further laboratory was set up in Scotland.

By 1960, the Animal Health Division successfully completed its programme to eradicate tuberculosis from the British cattle population. During the 1960s, the new challenges to animal health presented by the changing methods in livestock production gave extra impetus to the fight against animal disease. The intensive rearing of calves, for example, led to an increasing incidence of respiratory disease. This unforeseen consequence of new methods of animal husbandry required urgent bacteriological and virological investigation. Similarly, the development of artificial insemination and its widespread use in cattle presented new problems for animal disease control. The movement of inseminators between farms and the importation of and trade in semen presented new hazards for the transmission of infectious animal disease. In 1955, legislative measures were taken to prohibit importation of foreign animal semen without a licence and the Animal Health Division took steps to screen the health of bulls.

In addition to the Central Veterinary Laboratory, the Veterinary Investigation Service was established in 1922, with regional centres throughout Britain. This service essentially provided local laboratory facilities on which practising veterinarians could rely to help diagnose and investigate animal health problems in the herds that they attended. Initially, the Veterinary Investigation Centres were attached to agricultural colleges, but in 1946 they were incorporated into the Ministry of Agriculture, Fisheries and Food. The interaction between local veterinarians, veterinary investigation officers, the researchers of the Central Veterinary Laboratory and legislators in fact provides a good illustration of the increasing complexity of animal disease control during the twentieth century. At the grassroots level, practising veterinary surgeons observe and report the local incidence of animal disease.

This in turn is investigated by veterinary investigation officers who conduct field investigations and epidemiological studies to gauge the extent of animal health problems. In the meantime, the Central Veterinary Laboratory attempts to understand the aetiology of the disease and explores how it may effectively be controlled. These combined investigations may in turn lead to the inception of statutory control measures, vaccination, test and slaughter programmes to contain and eradicate animal disease. In Britain, this kind of cooperation eventually led to the eradication of brucellosis and is today responsible for the investigation and extensive monitoring of new pathogens and diseases, *e.g.* bovine spongiform encephalopathy. In October 1995, the Central Veterinary Laboratory and the Veterinary Investigation Service merged to form the Veterinary Laboratories Agency in a further bid to improve the collaboration necessary to ensure animal disease control and the provision of advice to the Ministry of Agriculture, Fisheries and Food and other veterinary health workers.

Throughout the twentieth century, the necessity for collaboration and cooperation has extended far beyond the activities of local and nationally organised veterinary organisations. International cooperation on matters of veterinary public health first began during the nineteenth century as a result of the devastation caused by the cattle plague. In 1924, the *Office International des Epizooties* was established in Paris. This intergovernmental organisation initially sought to promote and coordinate research into the control and surveillance of animal disease, and has remained active on this front to this day; its original membership of 28 having gradually been expanded to some 144 member states.

During the post-war period, additional international fora for dealing with the economic and health problems associated with animal disease were institutionalised. In Europe, the European Union—and its former incarnations, *i.e.* the EEC and EC—have provided the main political arena within which European governments can jointly discuss and legislate on matters pertaining to veterinary public health and animal welfare. The European Economic Community (EEC) was originally established in 1958- following the signing of the Treaties of Rome -with a view to achieving economic and political union amongst its member states, namely Belgium, Germany, France, Italy, Luxembourg and the Netherlands. In 1973, Denmark, Great Britain and Ireland became members of the EEC. In later years, Austria, Greece, Finland, Spain, Sweden and Portugal joined the ranks of states seeking European economic union; the organisation is set to expand yet further in future as former Eastern Bloc countries vie for membership.

Since its inception, the European Union has had a critical impact on agricultural practice and the trade in live animals and animal produce within Europe. The Common Agricultural Policy was instituted during the early 1960s with a view to increasing agricultural productivity, ensuring a fair standard of living for the agricultural community, stabilising markets, ensuring food supplies and providing consumers with food at reasonable prices. It has been perhaps the most important Common Policy that operates within the Union and has served as an important precursor to the 'single market' that is today intended to guarantee the free movement of goods, services, capital and people between the fifteen member states.

The Common Agricultural Policy greatly influenced the economics of livestock farming within Europe and initially encouraged farmers to intensify production. Ultimately this led to massive overproduction and the creation of huge surpluses, which in turn depressed market prices. 'Butter and beef mountains' came to symbolise the crisis in European agriculture; livestock numbers needed to be reduced and milk quotas were introduced. The Common Agricultural Policy has consequently undergone significant reforms and today it even encourages the *extensification* of production methods, particularly in the beef sector.

Free trade between member states remains at the top of the European Union's agenda, and trade naturally includes the trade in livestock and animal produce. As the epizootics of the past clearly evidenced, the movement of animals across national borders presents potentially great hazards to both public and animal health. Each member state has its own internal legislation and procedures regarding the control and prevention of livestock disease; the efficacy of which is entirely dependent on the efficiency of and cooperation between livestock producers, veterinarians and governmental agencies. However, membership of the European Union has required that each country's system of animal disease control be brought in line with those of other member states, in order to ensure a harmonisation and uniformity of disease prevention. Moreover, a system of veterinary inspection and veterinary and live-stock regulations have necessarily been introduced in an attempt to ensure the health of animals and the safety of meat that are crossing national borders.

With regard to livestock, each animal that is traded between member states, or enters the Union from a non-member state, must be accompanied by a veterinary health certificate that has been signed by an official veterinarian. Two types of certificates exist: one for animals intended for breeding and production; and another for animals intended for slaughter. The health requirements for breeding animals are much more stringent than for those destined for slaughter. These certificates detail the animal's origins, destination, means of transport, age, breed, type, means of identification and whether they passed through a market or assembly point. Such certificates are valid for ten days after the animal has been loaded for transport. In this way, diseased animals can be traced along with those that they came into contact with during transport.

The health certificate states that the animal has been examined and has been passed as healthy by an official veterinarian and that it originates from a herd or region free from specified diseases. Animals may have to undergo further tests depending on their origin, type and destination. Traded livestock must meet a number of welfare considerations determined on a Union-wide basis. Likewise, the rules for the trade in meat have been harmonised throughout the European Union, in order that animal flesh is produced, preserved and transported according to rules of hygiene identical to every member state. Certificates issued by official veterinarians at the place of meat production provide a guarantee that the meat is fit for consumption and serve as a valid transport document which testifies to the safety of the product. Produce from outside the European Union is subject to similar checks at border inspection posts and is required to meet the standards deemed acceptable within the Union. The veterinary regime has thus been further consolidated on an international level and has led—at least in principle—to the standardisation of the control of animal disease, welfare and health care and the quality of meat and other foods of animal origin within Europe. The legislators in Brussels have ensured that member states adopt uniform policies with regard to livestock production. European legislation, therefore, has increasingly come to determine the nature of national legislation. For example, since 1988, EU countries have maintained a ban on all growth-stimulating hormones, although such products are still widely employed in the USA. Attempts to eliminate the illegal trade in such hormone products even led to the murder of a Belgian veterinary inspector in 1995. An illustration of the impact of the EU's veterinary regulations on member states can also be provided by events surrounding the bovine spongiform encephalopathy (BSE) crisis. The severity of the legislation has greatly disabled the British livestock and meat trade and has also had significant repercussions for other countries.

It seems likely that the public health issues and trade problems surrounding BSE will continue to feature prominently on the European economic, agricultural and veterinary agenda in Brussels in the future. Similarly, the way the 1997 outbreak of classical swine fever in the Netherlands has been dealt with has to a great extent been affected by EU regulations. While vaccination may provide the key to ending the crisis, the EU has committed itself to a policy of eradication of classical swine fever and, at present, forbids the use of a vaccine which may be effective against *the* disease. The Dutch pig industry has been crippled by this outbreak, export embargoes have been instituted and over three million pigs have been destroyed, not necessarily because they had contracted the disease, but to prevent overcrowding of pig stalls.

While animal disease control and prevention measures have become increasingly stringent throughout Europe during the twentieth century, leading to the virtual eradication of the majority of animal plagues that decimated livestock and played havoc with the European economy in centuries past, we cannot forget that the modern global economy and the routine movement of individuals—both human and animal—across and between continents also have a bearing on veterinary public health within the bounds of Europe.

Diseases such as tuberculosis and classical swine fever, for example, are known to have been transmitted to western European livestock from live animal imports or animal produce from Eastern Europe.

Similarly, imports from countries where animal disease control is not so advanced or strict may still pose a threat to public health since spores of diseases such as anthrax, which have more or less been eradicated in advanced Western societies, can still be introduced through the import of hides, animal foods etc. from regions where the disease still persists. It must also be noted that while a highly effective veterinary regime exists within Europe, veterinary policing is not always as efficient as it should be and regulations are not always well enforced.

Essentially, it only takes one weak link in the chain of interdependencies involved in the process of transforming animals into safe food to lead to inefficiencies and failures in animal disease control. Again, the recent BSE crisis provides a clear example of this. Inadequate compliance with and policing of the meat and bone meal bans, first introduced by the British government in 1988 to prevent the feeding of mammalian protein to ruminants, appear to have allowed the epidemic to continue, when it might otherwise have been curtailed. One may thus conclude that any animal disease control measures that the legislators introduce can only be as good as their policing.

The control of animal disease today is far from simply a local or regional problem, it has instead become a global one. The involvement of veterinary experts in the policy making and execution of the World Health Organisation (WHO) and the Food and Agriculture Organisation (FAO) certainly attests to this. However, although advanced industrialised societies have the animal disease problem more or less under control—producing significant animal protein surpluses in spite of continuing population growth—the (rural) populations of developing nations across the globe continue to suffer both the economic and health effects of animal disease. Furthermore, animal diseases can affect livestock populations and food production to such an extent that they do not even return a profit. Livestock disease therefore not only generates economic problems, but also wastes resources which could help to solve the problem of world food shortages.

While the focus of this chapter is far from global, it is still important to be aware of the fact that the relationship between humans, animals and the environment in which they all live is a continually changing one which is not confined within national borders. The complexity and scale of interdependencies between humans and other animals is such today that failure to keep disease-causing microorganisms adequately in check could be potentially devastating for both the physical and economic health of world populations. The necessity of an effective and globally-oriented veterinary regime in present-day society is, therefore, manifest.

## CROP AND LIVESTOCK PRODUCTION

Overall, 1992 was a poor agricultural year. After having stagnated in 1991, global agricultural production only rose by about 1 percent in 1992. Except for North America, Oceania and the Near East, where production recovered from shortfalls experienced the previous year, regional agricultural performances ranged from mediocre to weak. Developing countries' agricultural production rose by only 1.7 percent, about half the average growth rate of the previous ten years.

Among developing country regions, only the Near East achieved some gain in per caput agricultural production (1.6 percent), which was, however, inadequate to compensate for the region's losses of the previous year.

In both the Far East and Latin America and the Caribbean, gains in agricultural production were entirely eroded by population growth. Africa recorded a 6 percent fall in per caput agricultural production, largely on account of the drought that affected developing southern Africa. Agricultural production in North America recovered significantly (up by 7.5 percent, compared with -1 percent in 1991 and a yearly average growth of 1 percent during the previous ten years) and accounted for more than 80 percent of the world's total production increase in 1992. Production also rebounded in Oceania (3.6 percent, three times the trend rate) but remained virtually unchanged in Western Europe.

Eastern Europe and the former USSR recorded yet another year of production decline: it was the fifth in six years in the former USSR, bringing the cumulative decline since 1987 to nearly 15 percent, and the third consecutive decline in Eastern Europe, with the overall fall in this region between 1989 and 1991 being more than 18 percent. While in the former USSR the shortfall in 1992 was chiefly due to a contraction in livestock production, that in Eastern Europe mainly reflected drought-affected crops.

### Per caput food production in 1992

The overall poor agricultural production performance of developing country regions was also reflected at the country level. No less than two-thirds of all developing countries recorded stagnant or declining levels of per caput food production. As has so often been the case in the past, the highest concentration of poor performances was in Africa, where all but six countries suffered per caput production losses. In some cases these were catastrophic, all the more so since they followed similarly dramatic crop shortfalls in the previous year. This was the case of Mozambique, Somalia, the United Republic of Tanzania and Zimbabwe. The few bright spots in the region included Chad, Ghana and, more significantly - given the population involved - Nigeria. These countries achieved sizeable gains in per caput food production for the second consecutive year.

## GREEN MANURES AND COVER CROPS

Green manuring consists of incorporating into the soil a crop grown for the purposes of soil improvement. It is a practice with a long history. Green manuring has been ignored in recent years as a serious option for soil improvement because the traditional practice entailed planting a full-season cover crop. This removed the field from commercial production for a whole season. Interest has returned, however, since green manuring strategies have been combined with cover cropping schemes.

Cover cropping is growing a crop for the purpose of soil and nutrient conservation. It is a more contemporary concept than green manuring, in crop agriculture. The two concepts — cover cropping and green manuring— go well together, as most cover crops are easily used as green manures prior to the planting of a commercial crop. The combined benefits become economically feasible when the cover is grown during the off-season or inter-seeded with the main crop. It is made even more desirable when the cover crop includes nitrogen-fixing legumes.

### Manuring and Composting

Livestock manures are the most traditional and widely recognized organic fertilizers. Under ideal circumstances, livestock enterprises are integrated into the whole farm operation, and manuring becomes part of a closed system of nutrient recycling. This is still strongly encouraged in organic operations. In reality, however, crops and livestock production are often divorced from each other, and manures must be imported.

This has created some concerns in the organic community, since much manure is now generated by large, industrial agriculture feeding operations called CAFOs (Confined Animal Feeding Operations). Not only are there concerns about contaminants (heavy metals, antibiotics, pesticides, hormones) but many in the organic community also object to any "partnering" with this segment of conventional agriculture, which is considered at odds with the environmental and social values represented by organic farming.

Nonetheless, the National Organic Programme does not differentiate between CAFO and other livestock manure sources. However, the NOP regulations do require that livestock manure not contain any synthetic substances not included on the National List of synthetic substances allowed for use in organic crop production. Another issue that has grown up around manure use in organic farming relates to food safety.

At a time when concerns about microbial contamination are high, there are questions about the risks associated with manure use on food crops. A focus piece on the February 2000 television news programme 20/20 was especially controversial. The segment suggested that organic foods were more dangerous than other food products in the marketplace due to manure fertilization. The reporter ignored the fact that conventional farms also use manures. Were all the manure generated annually in the U.S. (about 1.4 billion tons) applied only to organic farm acreage (estimated at roughly 1.5 million acres in 1997), each acre would receive about 933 tons. Furthermore, certified organic producers have strict guidelines to follow in handling and applying manures. The National Organic Programme regulations require raw animal manure be incorporated into the soil not less than 120 days prior to the harvest of a product whose edible portion has direct contact with the soil surface or soil particles, and be incorporated into the soil not less than 90 days prior to harvest of a product whose edible portion does not have direct contact with the soil surface or soil particles.

One of the best means of handling manures is composting. Composting stabilizes the nutrients in manure, builds populations of beneficial organisms, and has a highly beneficial effect on soils and crops. Compost can be produced on-farm by a number of means. Additional products from composts, such as compost teas, have special applications in organic agriculture.

Human manures are expressly forbidden in certified organic production. This includes composted sewage sludge (also called "biosolids"). The organic community made its opinion on this quite clear when the USDA's first draft of the national rule proposed allowing the use of sludge in certified production. It was counted as one of the "big three" targets of protest, along with food irradiation and genetic engineering. The prohibition of biosolids would have been disconcerting to Albert Howard, who decried the failure of cities to return their organic wastes to the countryside. Such recycling was, in his mind, a key aspect of sustainability.

## CROP ROTATION WITH COVER CROPS

One of the biggest challenges of cover cropping is to fit cover crops into your current rotations, or to develop new rotations that take full advantage of their benefits. This chapter will explore some of the systems used successfully by farmers in different regions of the U.S. One might be easily adapted to fit your existing crops, equipment and management. Other examples may point out ways that you can modify your rotation to make the addition of cover crops more profitable and practical.

Whether you add cover crops to your existing rotations or totally revamp your farming system, you should devote as much planning and attention to your cover crops as you do to your cash crops. Failure to do so can lead to failure of the cover crop and cause problems in other parts of your system. Also remember that there is likely no single cover crop that is right for your farm. Before you start:

Review Benefits of Cover Crops and Selecting the Best Cover Crops for Your Farm. Decide which benefits are most important to you.

Read the examples below, then consider how these cover crop rotations might be adapted to your particular conditions.

Talk to your neighbours and the other "experts" in your area, including the contact people listed in Regional Experts.

Start small on an easily accessible plot that you will see often.

Be an opportunist—and an optimist. If your cropping plans for a field are disrupted by weather or other conditions outside of your control, this may be the ideal window for establishing a cover crop.

Consider using an early-maturing cash crop to allow for timely planting of the cover crop.

Cover crops can be used for feed. Consider harvesting or grazing for forage or alternative livestock such as sheep and goats.

## Cover Crops for Corn Belt Grain and Oilseed Production

In addition to providing winter cover and building soil structure, nitrogen (N) management will probably be a major factor in your cover crop decisions for the corn>soybean rotation. A fall- planted grass or small grain will scavenge leftover N from the previous corn or soybean crop. Legumes are much less efficient at scavenging N, but will add N to the system for the following crop. Legume/grass mixtures are quite good at both.

### *Corn>Soybean Systems*

Keep in mind that corn is a heavy N feeder, soybeans benefit little, if at all, from cover crop N and that you have a shorter time for spring cover crop growth before corn than before soybeans.

Precaution. If you use herbicides, be sure to check labels for plant-back or rotation intervals to ensure that your cover crop isn't adversely affected.

Cover crop features: rye provides winter cover, scavenges N after corn, becomes a long-lasting (6 to 12 week) residue to hold moisture and suppress weeds for your soybeans; hairy vetch provides spring ground cover, abundant N and a moderate-term (4 to 8 week) mulch for a no-till corn crop; field peas are similar to vetch, but residue breaks down faster; red clover is also similar, but produces slightly less N and has less vigorous spring regrowth; berseem clover grows quickly to provide several cuttings for high-N green manure.

In Zone 7 and warmer, you can grow a cover crop every year between your corn and full-season beans. Also, you can use wheat or another small grain to replace the cover crop before beans, in a three-crop, two-year rotation (corn>wheat>doublecrop beans). In all cases, another legume or a grass/legume mixture can be used instead of a single species cover crop. Where it is adapted, you can use crimson clover or a crimson/grass mixture instead of vetch.

In cooler areas, plant rye as soon as possible after corn harvest. If you need more time in the fall, try overseeding in rowed beans at drydown "yellow leaf" stage in early fall, or in early summer at the last cultivation of corn. Seeding options include aerial application where the service is economical, using a specialty high-body tractor with narrow tires, or attaching a broadcast seeder, air seeder or seed boxes to a cultivator.

***Precaution.*** Broadcast seeding of cover crops into standing crops is less dependable than other seeding methods. Success will depend on many factors, including adequate rainfall amount and distribution after seeding.

Kill the rye once it is about knee-high, or let it go a bit longer, killing it a couple of weeks before planting beans. Killing the rye with herbicides and no-tilling beans in narrow rows allows more time for cover crop growth, since you don't have to work the ground. If soil moisture is low, consider killing the rye earlier. Follow the beans with hairy vetch or a vetch/small grain mixture.

Legumes must be seeded at least 6 weeks before hard frost to ensure winter survival. Seed by drilling after soybean harvest, or by overseeding before leaf drop. Allow the vetch (or mixture) to grow as long as possible in spring for maximum N fixation.

Harvesting sweet corn, seed corn or silage corn opens a window for timely cover crop planting. Harvest or graze the small grain or legume/small grain mixture in spring if needed for feed.

In Pennsylvania, Ed Quigley seeds rye or spring oats after corn silage harvest. The oats can be cut for silage in fall if planted by early September. Rye can be made into rylage or sprayed before no-till corn the following spring.

Worried about planting your corn a bit late because you're waiting for your cover crop to mature? Research in Maryland, Illinois and elsewhere suggests that no-tilling corn towards the end of the usual window when using a legume cover crop has its rewards. The delay can result in greater yields than earlier planting, due to greater moisture conservation and more N produced by the cover crop, or due to the timing of summer drought. In Pennsylvania, however, delayed planting sometimes reduced corn yields following rye.

Check your state variety trial data for a shorter season corn hybrid that yields nearly as well as slightly longer season corn. The cover crop benefit should overcome many yield differences.

Worried about soil moisture? There's no question that growing cover crops may consume soil moisture needed by the next crop. In humid regions, this is a problem only in an unusually dry spring.

Time permitting, allow 2 to 3 weeks after killing the cover crop to alleviate this problem. While spring rainfall may compensate for the moisture demand of most cover crops by normal planting dates, cover crops can quickly dry out a field.

Later in the season, killed cover crop residues in conservation tillage systems can conserve moisture and increase yields.

In any system where you are using accumulated soil moisture to grow your cash crop, you need to be extra careful. However, as noted in this chapter and elsewhere in the book, farmers and researchers are finding that water-thrifty cover crops may be able to replace even a fallow year without adversely affecting the cash crop.

***Corn>Rye>Soybeans>Small Grain>Hairy Vetch.*** This rotation is similar to the corn>rye>soybeans rotation described above, except you add a year of small grains following the beans. In crop rotation research from different areas, many benefits accrue as the rotation becomes longer. This is because weed, disease and insect pest problems generally decrease with an increase in years between repeat plantings of the same crop.

Residue from small grains provides good organic matter for soil building, and in the case of winter grains, the plants help to prevent erosion over winter after soybeans loosen up the soil. The length of the growing season will determine how you fit in cover crops after full-season soybeans in the rotation. Consider using a short- season bean if needed in order to achieve timely planting after soybean harvest. Calculate whether cover crop benefits will compensate for a possible yield loss on the shorter season beans. If there is not enough time to seed a legume after harvest, use a small grain rather than no cover crop at all.

The small grain scavenges leftover N following beans. Legume cover crops reduce fertilizer N needed by corn, a heavy N feeder. If you cannot seed the legume at least six weeks before a hard frost, consider overseeding before leaf drop or at last cultivation.

***Precaution.*** Because hairy vetch is hard seeded, it will volunteer in subsequent small grain crops.

An alternate rotation for the lower mid-South is corn>crimson clover (allowed to go to seed) > soybeans > crimson clover (reseeded) > corn. Allow the crimson clover to go to seed before planting beans. The clover germinates in late summer under the beans. Kill the cover crop before corn the next spring. If possible, choose a different cover crop following the corn this time to avoid potential pest and disease problems with the crimson clover.

***Precaution.*** In selecting a cover crop to interseed, do not jeopardize your cash crop if soil moisture is usually limiting during the rest of the corn season! Banding cover crop seed in row middles by using insecticide boxes or other devices can reduce cover crop competition with the cash crop.

***3 Year: Corn>Soybean>Wheat/Red Clover.*** This well-tested Wisconsin sequence provides N for corn as well as weed suppression and natural control of disease and insect pests. It was more profitable in recent years as the cost of synthetic N increased. Corn benefits from legume-fixed N, and from the improved cation exchange capacity in the soil that comes with increasing organic matter levels.

Growers in the upper Midwest can add a small grain to their corn>bean rotation. The small grain, seeded after soybeans, can be used as a cover crop, or it can be grown to maturity for grain. When growing wheat or oats for grain, frost-seed red clover or sweet-clover in March, harvest the grain, then let the clover grow until it goes dormant in late fall. Follow with corn the next spring. Some secondary tillage can be done in the fall, if conditions allow. One option is to attach sweeps to your chisel plow and run them about 2 inches deep, cutting the clover crowns.

Alternatively, grow the small grain to maturity, harvest, then immediately plant a legume cover crop such as hairy vetch or berseem clover in July or August. Soil moisture is critical for quick germination and good growth before frost. For much of the northern U.S., there is not time to plant a legume after soybean harvest, unless it can be seeded aerially or at the last cultivation. If growing spring grains, seed red clover or sweet-clover directly with the small grain.

An Iowa study compared no-till and conventional tillage corn>soybean>wheat/clover rotation with annual applications of composted swine manure. Berseem clover or red clover was frostseeded into wheat in March. Corn and soybean yields were lower in no-till plots the first year, while wheat yield was not affected by tillage. With yearly application of composted swine manure, however, yield of both corn and soybean were the same for both systems beginning in year two of the 4-year study.

Adding a small grain to the corn>soybean rotation helps control white mold on soybeans, since two years out of beans are needed to reduce pathogen populations. Using a grain/legume mix will scavenge available N from the bean crop, hold soil over winter and begin fixing N for the corn. Clovers or vetch can be harvested for seed, and red or yellow clover can be left for the second year as a green manure crop.

Using a spring seeding of oats and berseem clover has proved effective on Iowa farms that also have livestock. The mix tends to favour oat grain production in dry years and berseem production in wetter years. Either way the mixture provides biomass to increase organic matter and build soil. You can clip the berseem several times before flowering for green manure.

Precaution. Planting hairy vetch with small grains may make it difficult to harvest a clean grain crop. Instead, seed vetch after small grain harvest. Be sure to watch for volunteer vetch in subsequent small grain crops. It is easily controlled with herbicides but will result in significant penalties if found in wheat grain at the elevator.

### *Full-Year Covers Tackle Tough Weeds*

***Trout Run, PA.***—Growing cover crops for a full year between cash crops, combined with intensive tillage, helps Eric and Anne Nordell control virtually every type of weed nature throws at their vegetable farm—even quack-grass. They are also manipulating the system to address insects. The couple experimented with many different cover crops on their north-central Pennsylvania farm while adapting a system to battle quackgrass. Originally modelled on practices developed on a commercial herb farm in the Pacific Northwest, the Nordells continue to make modifications to fit their ever-changing conditions. In the fallow year between cash crops, the Nordells grow winter cover crops to smother weeds and improve soil.

Combined with summer tillage, the cover crops keep annual weeds from setting seed. Cognizant of the benefits of reduced tillage, they continue to modify their tillage practices—reducing tillage intensity whenever possible. Regular use of cover crops in the year before vegetables also improves soil quality and moisture retention while reducing erosion.

"Vegetable crops return very little to the soil as far as a root system," says Eric, a frequent speaker on the conference circuit. "You cut a head of lettuce and have nothing left behind. Growing vegetables, we're always trying to rebuild the soil."

Continual modification to their system is the name of the game. When they set up their original 4-year rotation in the early 80's, tarnished plant bugs were not an issue on their farm, but in the 90's they became a major problem in lettuce. The problem—and the solution!—was in their management of yellow sweetclover.

In their original rotation, sweetclover was overseeded into early cash crops such as lettuce. After overwintering, the sweetclover was mowed several times the following year before plowing it under and planting late vegetable crops. When the tarnished plant bugs began moving in—possibly attracted by the flowering sweetclover—the Nordells realized that mowing the sweetclover caused the plant bugs to move to the adjacent lettuce fields. It was time to change their system.

Fully committed to the use of cover crops, they first tried to delay mowing of the sweetclover until after lettuce harvest. Eventually, they decided to revamp their clover management completely. They now plant sweetclover in June of the second or fallow year of the rotation. This still gives the sweetclover plenty of time to produce a soil-building root system before late vegetables. It flowers later, so they are no longer mowing it and forcing tarnished plant bugs into lettuce fields.

Yellow blossom sweetclover—one of the best cover crop choices for warm-season nitrogen production—puts down a deep taproot before winter if seeded in June or July, observes Eric. "That root system loosens the soil, fixes nitrogen, and may even bring up minerals from the subsoil with its long tap root."

Originally part of their weed management programme, Eric points out that the clover alone would not suppress weeds. It works on their farm because of their successful management efforts over a decade to suppress overall weed pressure using intensive tillage, crop rotation and varied cover crops. The same concept applies to the tarnished plant bug. Never satisfied with a single strategy rather than a whole-system approach, the Nordells also began interseeding a single row of buckwheat into successive planting of short term cash crops like lettuce, spinach and peas.

The idea was to create a full-season insectiary in the market garden, moderating the boom and bust cycle of good and bad insects. They also hoped that the buckwheat would provide an alternate host for the plant bugs. The strategy seems to be working. Data collected as part of a research project with the Northeast Organic Network (NEON) found very few tarnished plant bugs in their lettuce but lots in the buckwheat insectiary.

The two pronged cover crop approach using buckwheat and a different management regime for sweetclover seems to be doing the trick. The next step, currently being evaluated, is to mix Italian ryegrass with the sweetclover to increase root mass and sod development between June planting and frost.

Rye and vetch are a popular combination to manage nitrogen. The rye takes up excess N from the soil, preventing leaching. The vetch fixes additional nitrogen which it releases after it's killed the following spring. With the August seeding, the Nordells' rye/vetch mixture produces" a tremendous root system" and much of its biomass in fall.

The Nordells plow the rye/vetch mix after it greens up in late March to early April, working shallowly so as not to turn up as many weed seeds. They understand that such early kill sacrifices some biomass and N for earlier planting of their cash crop—tomatoes, peppers, summer broccoli or leeks—around the end of May.

# 6

# Agricultural Biotechnology, Genomics and Acid Soils

## AGRICULTURAL BIOTECHNOLOGY

Broadly speaking, biotechnology is any technique that uses living organisms or substances from these organisms to make or modify a product for a practical purpose. Biotechnology can be applied to all classes of organism - from viruses and bacteria to plants and animals - and it is becoming a major feature of modern medicine, agriculture and industry. Modern agricultural biotechnology includes a range of tools that scientists employ to understand and manipulate the genetic make-up of organisms for use in the production or processing of agricultural products.

Some applications of biotechnology, such as fermentation and brewing, have been used for millennia. Other applications are newer but also well established.

For example, micro-organisms have been used for decades as living factories for the production of life-saving antibiotics including penicillin, from the fungus *Penicillium,* and streptomycin from the bacterium *Streptomyces.* Modern detergents rely on enzymes produced via biotechnology, hard cheese production largely relies on rennet produced by biotech yeast and human insulin for diabetics is now produced using biotechnology.

Biotechnology is being used to address problems in all areas of agricultural production and processing. This includes plant breeding to raise and stabilize yields; to improve resistance to pests, diseases and abiotic stresses such as drought and cold; and to enhance the nutritional content of foods.

Biotechnology is being used to develop low-cost disease-free planting materials for crops such as cassava, banana and potato and is creating new tools for the diagnosis and treatment of plant and animal diseases and for the measurement and conservation of genetic resources. Biotechnology is being used to speed up breeding programmes for plants, livestock and fish and to extend the range of traits that can be addressed. Animal feeds and feeding practices are being changed by biotechnology to improve animal nutrition and to reduce environmental waste. Biotechnology is used in disease diagnostics and for the production of vaccines against animal diseases.

Clearly, biotechnology is more than genetic engineering. Indeed, some of the least controversial aspects of agricultural biotechnology are potentially the most powerful and the most beneficial for the poor. Genomics, for example, is revolutionizing our understanding of the ways genes, cells, organisms and ecosystems function and is opening new horizons for marker-assisted breeding and genetic resource management. At the same time, genetic engineering is a very powerful tool whose role should be carefully evaluated. It is important to understand how biotechnology - particularly genetic engineering - complements and extends other approaches if sensible decisions are to be made about its use.

This chapter provides a brief description of current and emerging uses of biotechnology in crops, livestock, fisheries and forestry with a view to understanding the technologies themselves and the ways they complement and extend other approaches.

It should be emphasized that the tools of biotechnology are just that: tools, not ends in themselves. As with any tool, they must be assessed within the context in which they are being used.

The Convention on Biological Diversity (CBD) defines biotechnology as: "any technological application that uses biological systems, living organisms, or derivatives thereof, to make or modify products for specific use". This definition includes medical and industrial applications as well as many of the tools and techniques that are commonplace in agriculture and food production.

*The Cartagena Protocol on Biosafety defines "modern biotechnology" more narrowly as the application of:*

- In vitro nucleic acid techniques, including recombinant deoxyribonucleic acid (DNA) and direct injection of nucleic acid into cells or organelles, or
- Fusion of cells beyond the taxonomic family, that overcome natural physiological reproductive or recombination barriers and that are not techniques used in traditional breeding and selection.

Recombinant DNA techniques, also known as genetic engineering or (more familiarly but less accurately) genetic modification, refer to the modification of an organism's genetic make-up using transgenesis, in which DNA from one organism or cell (the transgene) is transferred to another without sexual reproduction.

Genetically modified organisms (GMOs) are modified by the application of transgenesis or recombinant DNA technology, in which a transgene is incorporated into the host genome or a gene in the host is modified to change its level of expression. The terms "GMO", "transgenic organism" and "genetically engineered organism (GEO)" are often used interchangeably although they are not technically identical. For the purposes of this report they are used as synonyms.

## Understanding, Characterizing and Managing Genetic Resources

*Table. An Agricultural Technology Timeline*

| Technology | Era | Genetic Interventions |
|---|---|---|
| **Traditional** | About 10 000 years BC | Civilizations harvested from natural biological diversity, domesticated crops and animals, began to select plant materials for propagation and animals for breeding |
| | About 3 000 years BC | Beer brewing, cheese making and wine fermentation |
| **Conventional** | Late nineteenth century | Identification of principles of inheritance by Gregor Mendel in 1865, laying the foundation for classical breeding methods |
| | 1930s | Development of commercial hybrid crops |
| | 1940s to 1960s | Use of mutagenesis, tissue culture, plant regeneration. Discovery of transformation and transduction. Discovery by Watson and Crick of the structure of DNA in 1953. Identification of genes that detach and move (transposons) |
| **Modern** | 1970s | Advent of gene transfer through recombinant DNA techniques. Use of embryo rescue and protoplast fusion in plant breeding and artificial insemination in animal reproduction |
| | 1980s | Insulin as first commercial product from gene transfer. Tissue culture for mass propagation in plants and embryo transfer in animal production |
| | 1990s | Extensive genetic fingerprinting of a wide range of organisms. First field trials of genetically engineered plant varieties in 1990 followed by the first commercial release in 1992. Genetically engineered vaccines and hormones and cloning of animals |
| | 2000s | Bioinformatics, genomics, proteomics, metabolomics |

Farmers and pastoralists have manipulated the genetic make-up of plants and animals since agriculture began more than 10 000 years ago. Farmers managed the process of domestication over millennia, through many cycles of selection of the best adapted individuals. This exploitation of the natural variation in biological organisms has given us the crops, plantation trees, farm animals and farmed fish of today, which often differ radically from their early ancestors. The aim of modern breeders is the same as that of early farmers - to produce superior crops or animals.

Conventional breeding, relying on the application of classic genetic principles based on the phenotype or physical characteristics of the organism concerned, has been very successful in introducing desirable traits into crop cultivars or livestock breeds from domesticated or wild relatives or mutants. In a conventional cross, whereby each parent donates half the genetic make-up of the progeny, undesirable traits may be passed on along with the desirable ones, and these undesirable traits may then have to be eliminated through successive generations of breeding. With each generation, the progeny must be tested for its growth characteristics as well as its nutritional and processing traits.

Many generations may be required before the desired combination of traits is found, and time lags may be very long, especially for perennial crops such as trees and some species of livestock. Such phenotype-based selection is thus a slow, demanding process and is expensive in terms of both time and money. Biotechnology can make the application of conventional breeding methods more efficient.

Spontaneous mutations are the "natural" motor of evolution, and the resource into which breeders tap to domesticate crops and to "create" better varieties. Without mutations, there would be no rice, or maize or any other crop. Starting in the 1970s, the International Atomic Energy Agency (IAEA) and FAO sponsored research on mutation induction to enhance genetic improvement of food and industrial crops for breeding new improved varieties. Induced mutations are brought about by treating plant parts with chemical or physical mutagens and then selecting for desirable changes - in effect, to mimic spontaneous mutations and artificially broaden genetic diversity. The precise nature of the mutations induced has generally not been a concern irrespective of whether the mutant lines were used directly or as sources of new variation in cross-breeding programmes.

Induced mutation to assist breeding has resulted in the introduction of new varieties of many crops such as rice, wheat, barley, apples, citrus, sugar cane and banana (the FAO/IAEA Mutant Varieties Database lists more than 2 300 officially released varieties). The application of mutation induction to crop breeding has translated into a tremendous economic impact on agriculture and food production that is currently valued in billions of US dollars and millions of hectares of cultivated land. Recently, mutation techniques have undergone a renaissance, expanding beyond their direct use in breeding into novel applications such as gene discovery and reverse genetics.

## SOCIAL RESISTANCE TO AGRICULTURAL BIOTECHNOLOGY

Over the last two decades, the life sciences industry has made enormous investments in biotechnology research and development; thrown tremendous energy into getting its genetically engineered (GE) crops approved, patented, and commercialized; and lobbied U. S. farmers and food producers to use them. For their part, the U. S., British, and other governments that envision the biotechnology sector as the wave of the future, and as a means of augmenting their national competitiveness, have strongly supported the industry and its efforts to commercialize (and normalize) these new technologies.

They have devoted considerable sums of money to biotechnology research, taken significant strides to deregulate the industry, and sought to promote the spread of U. S.-style intellectual property rights in the World Trade Organization. The U. S. government in particular has also promoted the dissemination of agricultural biotechnology in developing countries through the U. S. Agency for International Development.

With so much economic and political muscle propelling them, it is not surprising that GE crops hit the ground running when they came onto the scene in the mid-1990s. But what is surprising is that the rapid growth in GE crop deployment has been matched by an equally remarkable (and perhaps historically unprecedented) proliferation of citizens' voices challenging the biotechnology industry on economic, environmental, cultural, and moral grounds. Indeed, long before transgenic crops made their way to the market, individuals and groups concerned about the dissemination of these new technologies were already questioning their safety, utility, and necessity.

In this chapter, we explore the character and impact of the new social movement against genetic engineering in agriculture (hereafter, the "antibiotech movement"), as it is unfolding in the North, which is the heart of the agricultural biotechnology industry and the main arena in which these technologies have thus far been deployed. We analyse the strategies social activists have employed in efforts to limit the deployment of biotechnology in food and agricultural production, and their effectiveness in achieving this goal.

This antibiotechnology activism, we contend, has turned what until recently looked like a done deal in the trajectory of agricultural industrialization—the shift from farming to biotechnology, as David Goodman and his colleagues have put it — into a moment of uncertainty and openness. Activism has created this uncertainty, we argue, by affecting the fortunes of the biotechnology industry and the political environment in which it is developing in three substantial ways.

First, activism has forced states to reconsider their liberal regulatory approaches to agricultural biotechnology, and to take steps to regulate the industry more seriously. Although pushing recalcitrant states toward stronger regulatory oversight remains difficult, there are signs that morestringent national-level regulation will be forthcoming in many countries, including the United States. Second, antibiotech activism is helping to spawn new global regulatory regimes for genetically engineered organisms, such as the Cartagena Protocol on Biosafety and new Codex Alimentarius Commission standards on bioengineered foods.

Although social activists are not the main actors negotiating these regimes, they have helped ensure that their content moves the world toward more substantive regulation of these organisms. Third, the antibiotechnology movement has imposed important direct and indirect costs on the firms developing this technology. In addition, organized social activism has moved the issue of agricultural biotechnology out of relative obscurity, and out of the hands of a small number of corporate and state actors, into the public arena, where it is being debated by a broader spectrum of society.

In making this argument, we want to emphasize two points with respect to the recent sociological (as well as popular) literature on the industrialization of agriculture, of which biotechnology can be seen as the latest manifestation. First, in contrast to those who suggest that the events of the past several decades have provided transnational corporations with virtually unchallenged power to determine the future of world agriculture, we argue that social activists and their organizations are critical actors that have been overlooked by analysts of world-food-system restructuring.

Without exaggerating their power with respect to large corporations and powerful states, we argue that social activists are playing an underappreciated role in shaping patterns of food production, consumption, and distribution. In the case considered here, antibiotechnology activists have placed new demands on biotechnology firms and, even more significantly, on states, demands that have forced them to abandon certain paths and move in new directions.

Second, we want to distinguish ourselves from those who see consumer resistance as the primary basis for the recent change in the political and economic climate surrounding agricultural biotechnology. Although we do not mean to ignore the influence of consumer opposition, we hold that it is social-movement mobilization that has made consumers aware of genetic engineering of the food supply and that has motivated consumers to take a stance. Even in Europe, where consumer opposition has been energized by food and health scares such as mad cow disease in Britain, AIDS-tainted blood in France, and a dioxin scare in Belgium, it is activist nongovernmental organizations (NGOs) that have made agricultural genetic engineering into a social issue. We begin by providing a sketch of what the antibiotech movement looks like and who it comprises.

We then discuss the three main arenas in which the movement has concentrated its efforts in the North—the domestic regulatory sphere, the international regulatory sphere, and the corporate sphere—and evaluate its effectiveness. Finally, we consider the strengths and limitations of this type of activism.

### *Regulating beyond the Nation-State*

Activists have also sought to slow the biotechnology train by helping to create supranational regulatory regimes that govern the international deployment of GEOs in food and agriculture production. This strategy is reflected in the movement's efforts to create an international protocol (known as the Cartagena Protocol on Biosafety) regulating trade in transgenic organisms, as well as in its ongoing efforts to influence the work of the Codex Alimentarius Commission, the UN organization that sets international standards for food health and safety.

The idea for an international biosafety protocol was born during the negotiation of the UN Convention on Biological Diversity in 1992. During these negotiations, participants from the global South pointed out that their rich stores of biological diversity were uniquely vulnerable to the introduction of genetically engineered organisms. For this reason, Southern participants saw an urgent need for a protocol to oversee international trade in GEOs, which threatened to increase dramatically in the coming years.

But the United States and a handful of its allies (known collectively as the Miami Group) perceived such a protocol as a threat to the "right" of Northern farmers and corporations to produce and sell their goods, including genetically engineered agricultural commodities, worldwide and without prejudice. The group thus did its best to prevent the negotiation of such an agreement.

Ultimately, the powerful Miami Group lost. Although the final agreement was not as strong or comprehensive as most of the South wanted, it was broadly perceived by Southern negotiators and the antibiotech movement as a victory, since both groups had been seeking at least some form of control over GEO trade. Moreover, the fact that the regulatory regime would operate under the auspices of the relatively inclusive and democratic UN was perceived as a significant achievement, as the Miami Group had made a preemptive move to shift the onus of regulation to the clearly undemocratic World Trade Organization (WTO).

Although many factors contributed to the successful negotiation of the Cartagena Protocol in February 2000, the role of social activists was crucial. Prior to the negotiations, European activists and publics had put tremendous pressure on their governments to take genetic engineering seriously and to adhere to the "precautionary principle".

This pressure induced the European Union to dispatch a number of high-level ministers to the negotiations, sending a clear signal to other parties (including the United States) that Europe intended to be responsive to its citizens.

Representatives from a wide range of NGOs, from the North and South, also attended the meetings in person and used a variety of tactics to pressure delegates to reach a meaningful agreement. The NGOs engaged in a division of labour, sending some members into the meetings as formal observers and spokespersons and others onto the streets to demonstrate. Activist groups organized all-night vigils in the freezing Montreal weather, trotting out one colourful image after another to keep the media's attention.

Activists on the inside also worked around the clock, lobbying delegates and disseminating up-to-the-minute economic, legal, and scientific information. Shortly after the negotiations began, the activist network released a position statement, signed by twenty-five influential NGOs from around the world, that demanded a strong biosafety protocol and admonished the Miami Group for obstructing the negotiations. Activist organizations also sought to shore up the power of Southern delegations. Once it became clear that the United States had no interest in listening to the activists' position, they focused their efforts on forming a transnational alliance with state actors from the main negotiating bloc for the South, known as the LikeMinded Group. The activists supplied these Southern delegates with information, helped them assess proposed text, and made concrete recommendations about accepting, rejecting, or changing specific language. They also brought environmental scientists to the meetings to give public presentations, and invited the delegates to attend.

Reflecting on the negotiations, Tewolde Egziabher, the chief negotiator from the South, highlighted the importance of the activists' support when he explained how the Like-Minded Group, composed largely of African countries, managed to achieve anything against foes as powerful as the United States and its allies. "We had friends," Egziabher wrote.

Africa is financially so poor that the African Group would not have functioned ... without friends. But we soon made friends who filled in our gaps.... the Third World Network gave us critically needed assistance, and facilitated critically needed interactions, both South-South and South-North. African telecommunications are so poor that had it not been for the Gaia Foundation of London acting as an information relay station, we could not have been effective. And had our many, many other friends all over the world not helped, we would not have managed to stay as informed and as effective as we did.

As one observer put it, the activists "were the eyes watching for society at large, calling delegates on the carpet if they didn't remember that they have citizens back home who might not be happy if they do certain things.... They were the watchdogs."

Many of the same organizations are playing a similar watchdog role in the context of the Codex Alimentarius Commission. In 1999, the Ad Hoc Intergovernmental Task Force on Foods Derived from Biotechnology was set up within the Codex to develop standards and guidelines for foods derived from biotechnology.

The definition of these standards and guidelines—which include risk-assessment criteria and labeling requirements and which recognize the role of the precautionary principle in food safety standard setting—lies at the heart of the political struggles over the task force's work. Activists have followed the committee's deliberations closely, studying and critiquing the official positions taken by governments and their representatives and pressuring governments and the task force itself to try to influence its decisions. A central goal of these activist groups is to ensure that task force decisions do not contradict or override the biosafety protocol's provisions.

Activists are of course not the only ones trying to affect these decisions; every group with a strong interest in the Codex's decisions has sought to get its voice heard. In fact, one could argue that the activists' efforts are simply aimed at counteracting the influence of far more powerful Northern agribusiness interests.

Generally speaking, however, activists' efforts to establish and shape new international regulatory regimes for genetically engineered foodstuffs have helped to make decision making about agricultural biotechnology more democratic by ensuring that decisions about the deployment of biotechnology are not left entirely up to those with the most to gain from promoting these products and the least to lose if the products have negative economic or environmental consequences in importing nations.

### *Putting the Industry on the Defensive*

Social activism has not only forced the issue of biotechnology firmly on to the regulatory agenda but also complicated the economic lives of the biotechnology corporations. In recent years, church groups, consumers, farmers and farmers' associations, chefs, food processors, food retailers, and professional environmental and food-policy activists have all put direct pressure on these corporations, forcing them to rethink their strategies and, in some cases, to change their behaviour.

Perhaps most striking about the experience of the biotechnology industry in the 1990s is how unprepared it was for the strong reaction to its new attempts to "improve" nature. As one economist dryly observed, "Monsanto didn't even know a train was *coming* before it got run over." This may explain in part why the industry was so slow to react to the barrage of criticisms, consumer campaigns, lawsuits, and other actions that have thrown this techno-scientific project into question. But another part of the explanation surely resides in the divergent character—and preparation—of the two adversaries in this intense struggle for public opinion. On one side is a group of scrappy and diverse activist organizations who can punch and jab faster than you can say "genetic engineering," and on the other side is an elephant tethered to a post by its size, its enormous and sunk costs, and its competitive character.

Moreover, activist organizations have spent the last twenty years in basic training for just such a battle, one that requires particular political, organizing, and issue-framing skills and occurs on many different fields simultaneously. By contrast, many life sciences corporations were so preoccupied with building up their strategic positions in this long lead-time, capital-intensive, and risky industry and with trying to control the regulatory agenda that they barely knew they had an adversary in the making.

From this position of strategic if not economic advantage, antibiotechnology activists and other concerned groups have launched an all-out offensive against firms situated at virtually every point along the commodity chain. They have instigated shareholder actions against publicly owned corporations, demanding that they disavow the use of genetically engineered inputs.

Activists have also exerted pressure through lawsuits, such as one filed in 1999 by several prominent antitrust lawyers against Monsanto on behalf of small farmers and farm groups. The suit, inspired by a coalition of environmental groups, accused the biotechnology giant of being at the centre of an international conspiracy to control a large part of the world's seed supply and of giving farmers "false and fraudulent guarantees about the safety and marketability of a new breed of bioengineered seeds".

Activists have also organized consumer campaigns to convince food processors, retailers, and restaurant chains to stop selling genetically engineered foods, under the assumption that the market signals would quickly travel back up the supply chain to exporters, farmers, and, ultimately, the life sciences companies.

In response to consumer activism, a number of major supermarket chains, particularly in Europe, started to remove GE products from their shelves as early as 1998, and some North American and Canadian chains are now following suit.

The Organic Consumers Association and the BioDemocracy Campaign have also targeted fifteen food companies and distributors—dubbed the "Frankenfoods Fifteen"— to get them to stop using GE crops in their products. In general, these campaigns have been surprisingly successful. Since 1998, some of the world's major food processors, including H. J. Heinz, Gerber, Bestfoods, Frito-Lay, Unilever, Seagram's, and Nestle, have publicly declared their products to be GE-free in some or all their markets. And in August 2000, a major biotechnology producer, Novartis, declared that it was eliminating GE ingredients from all of its consumer food products—in effect refusing to buy its own products.

Consumer pressure—or the fear of it—has affected the whole food system. In 1999, McDonald's, fearing a threatened boycott, decided not to use genetically engineered potatoes in its French fries. After similar decisions were taken by Wendy's and Burger King, three major potato processors— J. R. Simplot, McCain Foods, and Lamb-Weston—advised farmers they would not be buying Monsanto's genetically engineered NewLeaf potatoes the following year. In consequence, Monsanto closed the research lab in Maine where the NewLeaf was developed and ultimately, pulled the potato from the market.

In Europe, these campaigns have succeeded in part because they tap into a smoldering anger about U. S. food imperialism. This anger was graphically depicted in 1999 when activist farmers dumped four tons of GE grain in downtown London from a truck bearing a large sign that read "Tony, Don't

Swallow Bill [Clinton]'s Seed. " When the French farmer José Bové, who trashed several McDonald's restaurants in his country and became a kind of folk hero, was asked what motivated his extreme actions, he replied, "To fight against globalization and advance the right of people to eat as they see fit". In the United States, activists have gathered support for their corporate campaigns by appealing to people's health concerns.

One poster printed up by Greenpeace depicts a Kellogg's cereal box, with a caption reading "Kellogg's Frosted Fakes." The Kellogg's mascot Tony the Tiger also appears, saying "Frankenfood BAD!" In general, activists have portrayed the industry as greedy and irresponsible, more concerned with profits than with public health, the environment, or poor countries' food security.

U. S. markets in Europe, Japan, and other countries have been significantly affected by the negative reaction to GEOs in the food supply. Between crop years 1997–98 and 1998–99, for instance, European imports of U. S. corn fell precipitously, from 2 million to 137,000 tons. Worried about losing their export markets, some farmers are adopting a more cautious attitude toward new GE crop varieties, such as GE wheat. Gary Goldberg, CEO of the American Corn Growers Association, argues that farmers' reactions are motivated by their economic concerns: "For agricultural producers, this debate over GEOs is not a safety, environmental or health issue. It is an economic issue.... Simply put, can farmers afford to grow a crop that they may not have a market for?"

These pressures have made it difficult for biotechnology firms to proceed with business as usual. Evidence that the biotechnology industry has been hurt is apparent both in the defensive moves the big firms have made in the last few years and in the industry's all-out effort to redeem itself in the eyes of the public, its direct customers (farmers), and its investors. For years, the industry kept a low profile, quietiy developing and commercializing its new gene technologies, but that strategy has become impossible.

The industry effectively admitted as much in November 1999, when seven of the largest life sciences firms formed a new industry alliance, the Council for Biotechnology Information (CBI), to counter criticism from antibiotechnology organizations. "The protest industry has gone too far," Edward Shonsey, a chief executive at Novartis Seed, told a *New York Times* reporter in November 1999.

"They've crossed the boundaries of reasonableness, and now it's up to us to protect and defend biotechnology". Through the CBI, the industry launched a $50 million advertising campaign to win back control of the discourse about agricultural biotechnology. Running the ad campaign are several high-powered public relations firms.

Individually and collectively, the life sciences firms are seeking to depict themselves as firmly committed to improving the social condition. One DuPont television commercial identifies one of its future tasks as "finding food to fight breast cancer. " Another portrays a poor old woman carrying a cooking pot in an unnamed Southern country; the caption proclaims that biotechnology will be able to "make food grow where food can't".

The health and nutrition page of the CBI's Web site highlights four areas in which the industry is poised to improve the human condition: by producing foods that will deliver oral vaccines, reduce or eliminate allergies, be more nutritious, and last longer and taste better. In addition, the life sciences firms are undertaking highly publicized goodwill efforts in the battle to remake their image. In early April 2000, Monsanto announced that it had a complete sequencing of the first crop-plant genome (rice) and then made this information publicly available by donating it to the International Rice Genome Sequencing Project, a consortium of ten publicly funded genome centres based in Japan. A month later, AstraZeneca announced that it would help make a genetically engineered strain of "golden rice" freely available to the developing world.

But perhaps the most significant evidence that social activism and consumer concerns are hurting the biotechnology industry financially is the shake-ups that have occurred in the biotechnology stock market and the fortunes of individual companies. Monsanto's stock price fell by half in 1999 because of the company's troubles in the agricultural biotechnology arena. In March 2000, Monsanto was compelled to merge with Pharmacia & Upjohn, which then hived off the agricultural part of its operation and put it on the market.

## BIOTECHNOLOGICAL PRINCIPAL OF AGRICULTURE

Agricultural technology can be defined as biotechnology to include 'any technique that uses living organisms or processes to make or modify products, to improve plants or animals or to develop micro-organisms for specific uses- it focuses upon recombinant DNA and cell fusion technologies'. Longworth concurs with this definition, with the significant addition of tissue-culture techniques, an aspect of biotechnology which has great economic, and therefore social and political, potential impact.

Despite debates about whether these or any other definitions are adequate, they do touch on the main aspects of biotechnology which deserve to be considered. The aspect of this new biotechnology which most captures the imagination and stirs the greatest controversy is gene splicing or recombinant DNA techniques, which inspire researchers to consider the possibilities of producing reproducible animals and plants markedly different from current existing species, and referred to as transgenic species. Already in higher animals and plants recombinant DNA has produced transgenic forms which are being commercially exploited. Their agricultural significance is, however, so far limited, and commercial application is concentrated in highly profitable pharmaceutical and horticultural markets.

Genes have been introduced into several animal species which alter their protein synthesis to enable transgenic sheep to produce insulin in their milk, and rabbits to produce interferon. A completely different application has originated in Denmark for salmon, where it has proved possible to introduce germplasm which enables the salmon's physiology to handle heavy metals, which are normally toxic, so opening up new locations for farm fisheries. Much current research is directed to conferring disease immunity on animals, and holds out the prospect of widespread commercial application.

In plants one achievement has been the transfer of genetic resistance to antibiotics in the petunia, and another has been the introduction of storage-protein genes from French bean plants into tobacco plants. As yet commercial progress with recombinant-DNA technology in plants appears limited, but extensive opportunities beckon, particularly because plant research is less restricted by the ethical and animal-welfare concerns which apply to research on transgenic animals. The most important commercial developments based on gene splicing have so far occurred with genetically much simpler microorganisms, and it is with these that the greatest short-to mediumrun commercial potential lies.

Already genetically engineered micro-organisms are producing a variety of hormones, vaccines, enzymes and other proteins. Important examples are the production of insulin, vaccines for neo-natal diarrhoea in calves and piglets, and the bovine growth hormone BST identical to that produced in cows which can stimulate a 20 per cent increase in milk yield. BST is already a source of problems for legislators in the European Community, and has provoked strong reactions from the media and milk consumers. As far as larger animals, and cattle in particular, are concerned, it is developments in embryo transfer and in many processes for manipulating reproduction which hold out the prospect of continuing increases in yields, feed conversion efficiency and general economic efficiency.

Already, apparently over 1 per cent of dairy calves in the USA are from embryo transplants, despite the high costs still associated with this procedure. Widespread adoption of these sophisticated technologies would further distance livestock farming from its traditional rural simplicity, and from the natural mating of animals to produce offspring. It would place technical demands upon the operators which favour large-scale company farms capable of supporting a range of highly trained specialists. Longworth identifies new techniques of tissue and cell culture as having 'the potential for enormous advances in crop improvement in the next couple of decades'. Those techniques 'can both increase the genetic diversity and greatly increase selection efficiency', and they permit innumerable plants to be reproduced asexually from single cells or small pieces of tissue.

The particular technique known as 'callus culture' has been used for many years to clone highly valued horticultural plants such as orchids, and cloning of cuttings is widely practised for tree crops such as tea and palm oil as well as by millions of gardeners for garden plants. According to Longworth 'cell culture' of single cells has unexpectedly, and so far inexplicably, resulted in plants with different properties being regenerated from the same clump of parent tissue. Sugar cane, maize and potato plants regenerated in this way have been found which are resistant to important pathogens.

This application of cell culture with the capacity to generate vast numbers of seedlings rapidly has the potential to simplify greatly the hitherto labourious procedures of plant breeding and selection. However, in terms of current and immediate commercial importance, it is through micro-organisms that biotechnology has its greatest impact. Microbial fermentation processes have been of great commercial significance for centuries, for example in bread, wine and cheese production, and there is the prospect of considerable development.

Two recent examples of new processes indicate the sorts of impact that such developments can have.

In the late 1960s genetically engineered bacteria were developed which were able to digest corn starch to produce high-fructose corn syrup (HFCS) and which left as a residue corn gluten, which is now an important protein feed for livestock. HFCS has made considerable inroads in the United States and in Japan. It has largely replaced sugar as a sweetener in Coca Cola as well as many other food and drink products. This has helped depress sugar prices to Third World growers. In the European Community steps have been taken to prevent HFCS and other new powerful sweeteners from being produced and from undermining the market for domestically grown beet sugar, which is supported by the Common Agricultural Policy.

A second important application of microbial fermentation has been the production of ethanol from sugar cane in Brazil and from maize in the USA. The commercial viability of these processes is critically dependent upon the price of oil as the main non-renewable source of fuel, and, to date, massive subsidies have been required to maintain the Brazilian and USA ethanol programmes. Longworth states, however, that a new biotechnology, Sucrotech, is being patented, which will not only reduce the cost of producing ethanol from sugar cane, but will simultaneously produce fructose at a cost which will be competitive with HFCS and may thereby reclaim part of the sweetener market for sugar cane.

That such possibilities are in prospect indicates how volatile the future might be; biotechnology has tilted the competitive balance from sugar cane to maize, causing economic pressure and even disruption to sugar-cane-dependent economies and may in future switch it back again. In the long run the capacity to ferment fuels microbially from renewable agricultural feedstocks points to an important long-term reorientation of agriculture if non-renewable oil becomes uncompetitively expensive as a fuel for cars and feedstock for certain chemicals.

It suggests that eventually there will be an increased emphasis on agricultural production of industrial feedstocks at the same time as continually increasing food output will be required to feed the expanding world population. All this will require considerable increases in agricultural productivity, to which biotechnology will increasingly contribute, but it will at the same time impose great strains on the natural environment and the structure of agriculture.

## GENOMICS

The most significant breakthroughs in agricultural biotechnology are coming from research into the structure of genomes and the genetic mechanisms behind economically important traits. The rapidly progressing discipline of genomics is providing information on the identity, location, impact and function of genes affecting such traits - knowledge that will increasingly drive the application of biotechnology in all agricultural sectors. Genomics sets the foundation for post-genomics activities, including new disciplines such as proteomics and metabolomics to generate knowledge on gene and protein structure, as well as their functions and interactions. These disciplines seek to understand systematically the molecular biology of organisms for their practical use. A vast range of new and rapidly advancing technologies and equipment has also been developed to generate and process information about the structure and function of biological systems. The use and organization of this information is called bioinformatics. Advances in bioinformatics may allow the prediction of gene function from gene sequence data: from a listing of an organism's genes, it will become possible to build a theoretical framework of its biology. The comparison across organisms of physical and genetic maps and DNA sequences will significantly reduce the time needed to identify and select potentially useful genes.

Through the production of genetic maps that provide the precise location and sequences of genes, it is apparent that even distantly related genomes share common features. Comparative genomics assists in the understanding of many genomes based on the intensive study of just a few.

For instance, the rice genome sequence is useful for studying the genomes of other cereals with which it shares features according to its degree of relatedness, and the mouse and malaria genomes provide models for livestock and some of the diseases that affect them. There are now model species for most types of crops, livestock and diseases and knowledge of their genomes is accumulating rapidly.

### *DNA*

All living things are made up of cells that are programmed by genetic material called deoxyribonucleic acid (DNA). Only a small fraction of the DNA chain actually makes up genes, which in turn code for proteins, and the remaining share of the DNA represents non-coding sequences whose role is not yet clearly understood. The genetic material is organized into pairs of chromosomes. For example, there are five chromosome pairs in the much-studied mustard species Arabidopsis thaliana. An organism's entire set of chromosomes is called the genome. The Human Genome Sequencing Project has provided the agricultural research community not only with many spin-off technologies that can be applied across the board for all living organisms but also with a model for international collaboration in tackling large genome-sequencing projects for model plants such as Arabidopsis and rice.

Synteny describes the conservation or consistency of gene content and gene order along the chromosomes of different plant genomes. Until well into the 1980s we imagined that each crop plant had its own genetic map. Only when we were able to make the first molecular maps, using a technique called "restriction fragment length polymorphism" (RFLP), did it begin to dawn on us that related species had remarkably similar gene maps.

The early experiments demonstrated conservation over a few million years of evolution in syntenous relationships between potato and tomato in the broad-leafed plants and between the three genomes of bread wheat in the grasses. Later we were able to show that the same similarities held over the rice, wheat and maize genomes, which were separated by some 60 million years of evolution. The diagram summarizes this research and shows 70 per cent of the world's food linked in a single map. The 12 chromosomes of rice can be aligned with the ten chromosomes of maize and the basic seven chromosomes of wheat and barley in such a way that any radius drawn around the circles will pass through different versions, known as alleles, of the same genes. The discovery of synteny has had an enormous impact on the way we think about plant genetics. There are obvious applications for evolutionary studies; for example, the white arrows on the wheat and maize circles describe evolutionary chromosomal translocations that describe *Pooideae* and *Panicoideae* groups of grasses. There are great opportunities to predict the presence and location of a gene in one species from what we know from another. Now that we have the complete DNA sequence of rice we are able to identify and isolate key genes from large genome intractable species such as wheat and barley by predicting that the same genes will be present in the same order as in rice. Key genes for disease resistance and tolerance to acid soils have recently been isolated from barley and rye in this way.

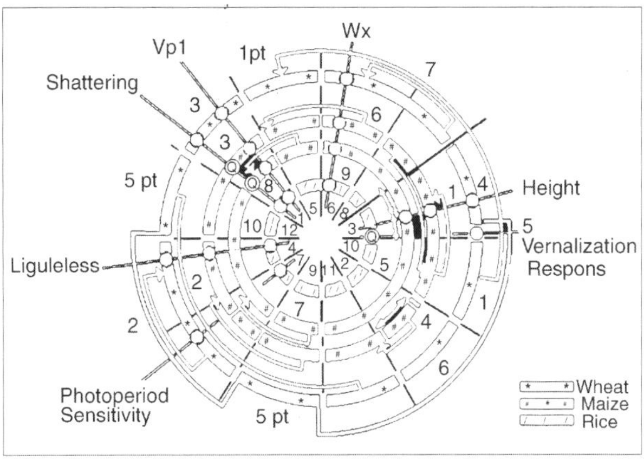


For practical plant breeding, knowledge of synteny allows breeders access to all alleles in, for example, all cereals rather than just the species on which they are working. A key first example of this is the transfer to rice of the wheat dwarfing genes that made the Green Revolution possible. In these experiments the gene was located in rice by synteny and then isolated and engineered with the alteration in DNA sequence that characterized the wheat genes before replacing the engineered gene in rice. This approach can be applied to any gene in any cereal, including the so-called "orphan crops" that have not attracted the research dollars that the big three - wheat, rice and maize - have over the past century. The main significance is, however, that we can now pool our knowledge of biochemistry, physiology and genetics and transfer it between crops via synteny.

## Molecular Markers

Reliable information on the distribution of genetic variation is a prerequisite for sound selection, breeding and conservation programmes. Genetic variation of a species or population can be assessed in the field or by studying molecular and other markers in the laboratory. A combination of the two approaches is required for reliable results. Molecular markers are identifiable DNA sequences, found at specific locations of the genome and associated with the inheritance of a trait or linked gene.

*Molecular markers can be used for:*

- Marker-assisted breeding,
- Understanding and conserving genetic resources
- Genotype verification. These activities are critical for the genetic improvement of crops, forest trees, livestock and fish.

## MARKER-ASSISTED BREEDING

Genetic linkage maps can be used to locate and select for genes affecting traits of economic importance in plants or animals. The potential benefits of marker-assisted selection (MAS) are greatest for traits that are controlled by many genes, such as fruit yield, wood quality, disease resistance, milk and meat production, or body fat, and that are difficult, time-consuming or expensive to measure. Markers can also be used to increase the speed or efficiency of introducing new genes from one population to another, for example when wishing to introduce genes from wild relatives into modern plant varieties. When the desired trait is found within the same species, it may be transferred with traditional breeding methods, with molecular markers being used to track the desired gene.

Pearl millet is a cereal grown for foodgrain and straw in the hottest, driest areas of Africa and Asia where rainfed and dryland agriculture are practised. It is similar to maize in its breeding behaviour. Traditional farmers' varieties are open-pollinated and out-breeding and thus continuously changing. Genetically uniform hybrid varieties have been developed that offer higher yield potential but are more vulnerable to a plant disease called downy mildew.

In India, pearl millet is grown on about 9 million ha and more than 70 per cent of this is sown to such hybrid cultivars. Since pearl millet hybrids first reached farmers' fields in India in the late 1960s, every variety that has become popular with farmers has ultimately succumbed to a downy mildew epidemic. Unfortunately, by the time the poorer farmers in a given region decide to adopt a particular variety, its days are usually numbered.

The International Crops Research Institute for the Semi-Arid Tropics (ICRISAT) wanted to reduce the risks associated with adoption of higher-yielding pearl millet hybrids and extend the useful economic life of these varieties, especially for poorer producers. Biotechnology helped us to achieve this. With tools from the John Innes Centre and support from the Plant Sciences Research Programme of the Department for International Development (DFID), we developed and applied molecular genetic tools for pearl millet. We mapped the genomic regions of pearl millet that control downy mildew resistance, straw yield potential, and grain and straw yield under drought stress conditions.

Then our millet breeders used conventional breeding and marker-assisted selection (MAS) to transfer several genomic regions conferring improved downy mildew resistance to the two elite inbred parental lines of popular hybrid HHB 67. We then used MAS to derive two new varieties - ICMR 01004 and ICMR 01007 - with two different downy mildew resistance gene blocks.

These varieties have performed as well or better than their parent lines for grain and straw yield, and are markedly improved for downy mildew resistance. They also retain several favourable traits, including 1 000-grain mass, panicle length, plant height and rust resistance. Hybrids based on crosses involving ICMR 01004 and ICMR 01007 have recently advanced to trials in the Indian states of Gujarat, Rajasthan and Haryana under the All India Coordinated Pearl Millet Improvement Project. This follows their successful evaluation in 2002, in which they exhibited marginal grain yield superiority and substantially better downy mildew resistance than HHB 67, while maintaining the early maturity that contributes to its popularity.

At least one of these two hybrids could be released as a replacement for HHB 67 before the latter succumbs (as it surely will) to a downy mildew epidemic. Because HHB 67 is so widely grown by poor farmers in India, if its timely replacement could prevent such an epidemic for even one year, the losses avoided would exceed the total value of research-funding support by DFID for the development and application of the molecular genetic tool kit for pearl millet (£3.1 million to date). All future benefits from this research by ICRISAT, its DFID-supported partners in the United Kingdom, and collaborating national programme partners in India can then be considered profits to society.

### Measuring and Conserving Genetic Diversity

The use of molecular markers to measure the extent of variation at the genetic level, within and among populations, is of value in guiding genetic conservation activities and in the development of breeding populations in crops, livestock, forestry and fisheries. Studies carried out using these technologies in fish and forest tree species have revealed high levels of genetic variation both among and within populations. Livestock species are characterized by a high degree of genetic variation within populations, whereas crops exhibit a higher degree of variation across species. Data from other approaches, for example field observation, often cannot provide such information or are extremely difficult to collect.

Molecular markers are increasingly used to study the distribution and patterns of genetic diversity. Global surveys indicate, for example, that 40 per cent of the remaining domestic livestock breeds are at risk of extinction. Most of these breeds are found only in developing countries, and there is often little knowledge about them or of their potential for improvement. They may contain valuable genes that confer adaptation or resilience to stresses, such as heat tolerance or disease resistance, that may be of use for future generations. Modern biotechnologies can help to counteract trends of genetic erosion in all food and agriculture sectors.

## AGRICULTURE ON ACID SOILS

Aluminium in acid soils limits plant growth on more than 30 per cent of all arable land, primarily in developing countries. There are two approaches to increasing crop production on acid soils. Lime can be added to the soil to increase the pH, but this is a costly, temporary measure. Alternatively, genetically improved cultivars, tolerant to aluminium, can be developed.

Existing wheat cultivars do not contain significant genetic variation for increasing aluminium tolerance. Improved tolerance will have to be introduced into wheat from the gene pools of related, more tolerant species. A genetic linkage map of wheat was developed using available markers for the existing aluminium-tolerance gene. Rye exhibits a fourfold increase in aluminium tolerance over wheat. Therefore, a rye gene controlling aluminium tolerance was characterized. Markers from wheat, barley and rice were used to establish a tight linkage, flanking the rye gene, and to construct a high-resolution genetic map. A potential candidate gene was used for root-gene-expression, time-course studies that showed expression in rye roots only under aluminium stress.

Targeting the aluminium tolerance gene is one example of using problem-based approaches to integrate molecular and breeding tools to improve wheat production. Using the genetic relationship (synteny) among the cereals to supply markers to identify and characterize value-added traits, complementary approaches for improved wheat production emerge. Breeders can use the markers flanking the rye gene in marker-assisted breeding programmes in areas where GMOs cannot be grown or where only conventional breeding tools are available.

In addition, these markers can be used for map-based cloning to isolate the gene in question for transgenic approaches to wheat improvement. Finally, the use of syntenous relationships offers the technology to manipulate many value-added traits for crop improvement in other species.

## Research

Researchers at Jawaharlal Nehru University in India have developed a genetically engineered potato that produces about one-third to one-half more protein than usual, including substantial amounts of all the essential amino acids such as lysine and methionine. Protein deficiency is widespread in India and potato is the staple food of the poorest people.

The "protato" was developed by a coalition of Indian charities, scientists, government institutes and industry as part of a 15-year campaign against childhood mortality.

The campaign aims to eliminate childhood mortality by providing children with clean water, better food and vaccines. The protato includes a gene from the amaranth plant, a high-protein grain that is native to South America and widely sold in Western health-food stores. The protato has passed preliminary field trials and tests for allergens and toxins. Final approval from the Indian Government is probably at least five years away.

Supporters such as Govindarajan Padmanaban, a biochemist at the Indian Institute of Science, argue that the protato can provide an important nutritional boost to children with little danger of allergy because potatoes and amaranth are both already widely consumed. There is also little threat to the environment because neither potatoes nor amaranth have wild relatives in India, and the protato does not involve any change in normal potato production practices.

Furthermore, because the protato was developed by public-sector scientists in India, there are no concerns about foreign corporate control of the technology.

*Given these benefits, Padmanaban commented:*

- "I think it would be morally indefensible to oppose it".

Opponents such as Charlie Kronick of Greenpeace argue that potatoes are naturally quite low in protein (about 2 per cent), so even a doubling of the protein content would make only a minute contribution to India's malnutrition problem. He claims that the effort to develop the protato was aimed more at gaining public acceptance of genetic engineering than at addressing the problem of malnutrition:

- "The cause of hunger isn't lack of food. It's lack of cash and of access to the food. Creating these GM crops is something to make them look attractive when actually the utility of eating them is very, very low. It's very difficult to see how this on its own will change the face of poverty".

*Breeding and Reproducing Livestock and Fish*

Biotechnology has long been a source of innovation in livestock and aquaculture production and processing and has had a profound impact on both sectors. Rapid advances in molecular biology and further developments in reproductive biology provide new and powerful tools for further innovation. Technologies such as genomics and molecular markers are valuable in understanding, characterizing and managing genetic resources in livestock and fisheries as well as in crops and forestry. Genetic engineering is also relevant in livestock and fisheries, although the techniques differ, and additional reproductive technologies are available in these sectors. This section describes the reproductive biotechnologies that are specific to the livestock and fisheries sectors.

The main objective of reproductive biotechnologies for livestock is to increase reproductive efficiency and rates of animal genetic improvement. The genetic improvement of locally adapted breeds will be important in realizing sustainable production systems within the broad spectrum of developing country production environments, and will probably best be realised by the strategic use of both non-genetic and genetic interventions. Reproductive biotechnology in fisheries presents opportunities to increase growth rates and improve the management of farmed species and to limit the reproductive potential of genetically engineered species.

As part of this country-driven strategy for the management of farm animal genetic resources, FAO invited 188 countries to participate in preparing the First Report on the State of the World's Animal Genetic Resources, to be completed before 2006. To date, 145 countries have agreed to submit country reports and 30 country reports have been received and analysed.

It is clear from these reports that artificial insemination (AI) is the most common biotechnology used by developing countries in the livestock sector. Many countries have requested training for the expansion of AI use, while expressing concerns that it has often been introduced without proper planning and may pose a potential threat to the conservation of local breeds.

Although the use of multiple ovulation followed by embryo transfer (MOET) is mentioned and the desire for its introduction or expansion expressed, no clear objectives for this technique are mentioned. All countries have expressed the wish to introduce and develop molecular techniques, often as a complement to phenotypic breed characterization.

Cryoconservation was identified as a priority by all countries and gene banks were recommended, but funding remains a major constraint. When animal GMOs are mentioned it is mainly to express the lack of proper regulations and guidelines for their eventual production, use and exchange. Some countries have expressed concerns that biotechnologies in the livestock sector should be, but are not always, pursued as an integral part of an overall genetic improvement strategy.

## Artificial Insemination and Multiple Ovulation/Embryo Transfer

Advances in artificial insemination (AI) and multiple ovulation followed by embryo transfer (MOET) have already had a major impact on livestock improvement programmes in developed countries and many developing countries because they speed up the process of genetic improvement, reduce the risk of disease transmission and expand the number of animals that can be bred from a superior parent - the male in the case of AI and the female in the case of MOET. They also increase the incentives for private research in animal breeding and significantly expand the market for improved parent stock.

The number of AIs performed globally during 1998 was over 100 million in cattle (primarily dairy cattle, including buffalo), 40 million in pigs, 3.3 million in sheep and 0.5 million in goats. These figures illustrate both the higher economic returns in dairy cattle and the fact that cattle semen is much easier to deep-freeze than semen from other animals. Although over 60 million cattle AIs were performed in South and Southeast Asia, fewer than 1 million were performed in Africa.

AI is only effective when the farm sector has access to considerably greater technical and institutional and other organizational capacity than when male animals are used directly for breeding purposes. On the positive side, farmers employing AI do not have to face the costs or hazards of rearing breeding males and can have access to semen from any part of the world.

Despite the widespread use of AI in developed countries and in many developing countries, including within more advanced smallholder systems, it is applied only on farms that practise intensive livestock management with high-value animals. This is clearly not because of technical problems with semen production and storage, as most procedures are now fully standardized and proven to be effective under tropical developing-country conditions. Rather, it is because of the many organizational, logistical and farmer-training constraints that influence the quality and efficiency of the technology.

MOET takes AI one step further, both in terms of genetic gains possible and level of technical capacity and organization required. MOET is one of the basic technologies for the application of more advanced reproductive biotechnologies such as cloning and transgenics. During 2001 the number of embryos transferred globally was 450 000, mainly in dairy cattle, with North America and Europe accounting for 62 per cent, followed by South America (16 per cent) and Asia (11 per cent). About 80 per cent of the bulls used in AI are derived from MOET. The main potential advantage of MOET for developing countries will be in the possibility of importing frozen embryos instead of live animals, for example in the establishment of nucleus breeding stocks of locally adapted genetic resources, with the related lower sanitary risks.

### *Chromosome-set Manipulation and Sex Reversal in Fish*

Controlling the sex and reproductive capacity of fish can be important for commercial and environmental reasons. One sex is often more desirable than the other; for example, only female sturgeon produce caviar and male tilapia grow faster than females. Sterility may be desirable when reproduction affects the taste of the product (*e.g.* oysters) or when farmed species (transgenic or not) might breed with wild populations. Chromosome-set manipulation and sex reversal are well-established techniques to control these factors. In chromosome-set manipulation, temperature, chemical and pressure shocks applied to fish eggs can be used to produce individuals that have three sets of chromosomes rather than the usual two. These triploid organisms generally do not channel energy into reproduction and thus are functionally sterile. Sex reversal can be accomplished by several methods including administering appropriate hormones. For example, genetically male tilapia can be turned into females through oestrogen treatments. These genetic males, when mated with normal males, produce a group of all-male tilapia.

### *Genetic Engineering in Livestock and Fish*

Genetic engineering in animals can be used to introduce foreign genes into the animal genome or, alternatively, to "knock-out" selected genes. The method most used at present involves direct microinjection of DNA into the pronuclei of fertilized eggs, but progress is being made with new approaches such as nuclear transfer and the use of lentiviruses as DNA vectors.

In the first genetic engineering experiments with farm animals, genes responsible for growth were introduced into pigs to increase growth and improve carcass quality. Current research efforts include engineering resistance to animal diseases, such as Marek's disease in poultry, scrapie in sheep and mastitis in cattle, and diseases that affect human health such as *Salmonella* in poultry. Other examples include increasing the casein content of milk and inducing the production of pharmaceutical or industrial chemicals in the milk or semen of animals. Although conceptually simple, the methods used to genetically engineer livestock require special equipment and considerable dexterity, and no agricultural applications have proved commercially successful thus far. Applications in the near future therefore seem to be limited to the production of transgenic animals for use in the production of industrial or pharmaceutical products.

Genetic engineering is an active area of research and development in aquaculture. The large size and hardy nature of many fish eggs allow them to be manipulated easily and facilitate gene transfer by direct injection of a foreign gene or by electroporation, in which an electric field assists gene transfer. Gene transfer in fish has usually involved genes that produce growth hormone and has been shown to increase growth rates dramatically in carp, salmon, tilapia and other species.

In addition, a gene from the winter flounder that produces an antifreeze protein was put into salmon in the hope of extending the farming range of the fish. The gene did not produce enough of the protein to extend the salmon's range into colder waters, but it did allow the salmon to continue growing during cold months when non-transgenic salmon would not grow. These applications are still in the research and development stage, and no transgenic aquatic animals are currently available to the consumer.

## FUTURE OF AGRICULTURAL BIOTECHNOLOGY

There are divergent views about the rate of uptake of new agricultural biotechnologies and their impact. Kalter and Tauer and the US Office of Technology Assessment anticipate comparatively rapid take-up, while others

(Buckwell and Moxey and Farrington do not foresee much impact until the next century, that is until ten to fifteen years have elapsed. The caution of the latter commentators is probably justified, for there are various hurdles to be jumped before biotechnologies can gain acceptance. One hurdle to be overcome is the economic one.

Self-evidently there must be some economic advantage to farmers or agro-industry from adopting the technology. Frequently what captures the scientific imagination turns out to be economically unviable in use, although it is only with use that efficiency can be improved and costs brought down. A more demanding test still for biotechnologies will be to obtain legislative approval and public acceptance. These are intimately connected.

In the case of machinery developments, provided safety standards are observed there are no legal impediments to developing new and improved machines, since there are no obvious problems of hazard to the public and therefore public acceptance. Chemical insecticide, fungicides and herbicides do, however, pose much greater problems because of concerns with toxicity, and elabourate testing and licensing procedures have evolved after a relatively haphazard set of procedures in the 1950s and 1960s, when widespread use was made of DDT and organo-phosphorous compounds without adequate recognition of their toxicity.

In significant part, because of the problems which have been experienced with agro-chemicals, the licensing of biotechnologies will inevitably be based on testing at least as stringent as that which now exists for chemicals. Indeed the licensing procedures are likely to be more stringent because of public and scientific concerns. Public acceptance of biotechnology in the food chain will be made more difficult by confusion about differences between biotechnologies. In the early 1980s there was a crescendo of concern about the use of steroids to promote faster liveweight gain in calves and cattle.

The public outcry eventually led to the banning of such hormones for meat animals; but the legacy is a profound distrust of and hostility to new products such as synthetic Bovine Somatrophin (BST), which has the capacity to increase the milk yield of those dairy cows which are 'relatively' deficient in what is a naturally occurring hormone. Although initial scientific results are favourable, and some minor doubts remain, the major influence leading the European Commission to impose a moratorium on the use of BST until the end of 1990 at least is concern about public perception; the Commission has stated 'It would be a serious setback to producers and to the Community's milk policy were present [positive] trends in consumption to be reversed as a result of adverse consumer reaction. With newly bioengineered plants and animals scientific and public concerns are emerging which are of a different order from those associated with previous biotechnology.

One relates to the possibility that crop failures may become more frequent if biotechnology leads to a reduction in genetic diversity in crops being grown; such a tendency has already been observed in relation to the Green Revolution technology for wheat. There are also concerns about upsetting the balance of nature in unforseen ways, akin to the unanticipated consequence of introducing rabbits into Australia and then trying to control these by introducing myxomatosis.

Thus questions arise such as what happens if herbicide resistance introduced into a commercial crop plant transfers itself by cross pollination into a closely related weed species, or indeed if the herbicide-resistant crop should colonize wild habitats. There are, of course, more lurid and absurd notions bandied about in the popular press which are similar in nature to the attacks made on Darwinism. The questions, both absurd and real, raised in relation to agricultural biotechnology will undoubtedly slow the rate at which it is adopted. Nevertheless instances of adoption are increasing: BST in the USA, a bioengineered baker's yeast in the UK, bioengineered sheep producing insulin, etc. As these become more widespread and increasingly affect lower-valued, bulk agricultural products, so the impacts of biotechnology on structural change will increase. The balance of economic power will switch increasingly to industry, to high-technology large farms and against smaller farmers in disadvantaged regions and countries. That, unfortunately, is a seemingly inevitable consequence of what we consider to be economic progress.

# 7

# Food Safety and GM Crops Policy

## FOOD SAFETY AND CONSUMER CHOICE

As of 2000, India had not yet developed an explicit consumer food safety policy toward GM crops. This partly reflected India's status as a low-income developing country where many poor consumers were still concerned as much about the availability or the price of food as about its exact characteristics or ingredients. Yet this also reflected India's official claim that GM foods were not yet being sold in the country. In this regard, India was in a similar position to Kenya. If the Government of India begins approving GM food crops (not just cotton) for commercial production, it may need to develop a more explicit food safety and consumer policy as an accompaniment.

Consumer protection laws have been slow to emerge overall in India, partly reflecting the nation's poverty yet also partly reflecting a large role played by the state itself in providing consumer goods through state owned industries. The state has historically been reluctant to sponsor a strong consumer protection movement partly for fear of liabilities to its own public industries. The notion of protecting consumers by giving them an "informed choice" has also been slow to develop in India, partly because India's citizens and its government officials have historically been comfortable with a more direct regulatory approach. Consumer advocates in India tend to focus more on issues of state regulatory failure rather than informed choice. In the food safety area, for example, the state is routinely held accountable by opposition parties or through the media for occasional failures to prevent the contamination or adulteration of processed and packaged foods. As noted above, a 1998 dropsy outbreak in Delhi linked to contamination of edible oils prompted a quick tightening of packaging rules. Partly because so much food consumption in India has traditionally been satisfied from unpackaged natural ingredients, consumer fears regarding adulteration of packaged foods remain understandably widespread. According to one survey, 60 percent of Delhi's population fears that branded milk could be adulterated, and hence unsafe to drink (Rao 1999).

Because India does not yet officially grow or import any GM foods, it has been able to continue operating within food safety policies that draw little or no distinction between GM and non-GM food ingredients. India's still operative 1954 Prevention of Food Adulteration Act predates the GM crop revolution and does not mention transgenic entities. In 1998, however, India made at least one adjustment to the GM crop revolution when it revised its biosafety approval guidelines to require that GM seeds, plants and plant parts be separately screened for toxicity and allergenicity (DBT 1998). By introducing a separate regulatory procedure for GM crops, this new RCGM guideline gives India a permissive rather than a fully promotional safety policy toward GM foods. The protocols published in the revised guidelines do not set a higher food safety standard because of the GM nature of the products; they were developed by RCGM through consultation with conventional industrial toxicologists and scarcely go beyond the testing that might be appropriate for judging the safety of pesticide residues.[6] Also, the new procedures allow for test results generated in other countries to be used in India, on the grounds that food safety testing does not have to be site specific. This means the data used by FDA to give approval to a GM food in the United States could presumably be re-used for the same purpose in India. The process thus can be a quick one, and RCGM officials have already indicated informally that in the GM mustard application (the first food crop application going through India's approval system) their review of the information submitted has already found no new food safety risks (Ghosh and Ramanaiah 2000).

Labeling policies in India have been a moot question until now because of the nation's nominal GM-free status and because most food consumption continues to be satisfied through home or street preparations of natural foods that are never packaged at all, let alone carefully labeled. Foods grown in India for export purposes must be labeled according to the policies of the importing countries, yet the nation's GM-free status has transformed this requirement into something of a competitive advantage for the country rather than a burden. Without any costly market segregation, all of India's soy or castor oilseed cake destined for Europe and Japan can be labeled "GM free."

Labeling requirements are likely to be imposed as soon as GM food crops (as opposed to fiber crops like cotton) are released by GEAC for commercial planting by India's own farmers. Transgenic pharmaceutical products in India that have been deregulated for sale be GEAC already must be labeled GM. Anticipating the GM labeling problem for foods, India's Ministry of Food Processing Industries has approached DBT seeking advice, and India's Ministry of Health and Family Welfare has approached both DBT and the Codex Alimentarius in Rome asking how the nation's 1954 Prevention of Food Adulteration Act can be adjusted to take GM ingredients into account. If and when GM foods do officially come onto the market in India, DBT can be expected to advocate labeling requirements that are permissive rather than precautionary, to avoid a need for full market segregation.

## PUBLIC INVESTMENTS IN RESEARCH

The Government of India, principally through its Department of Biotechnology, has for more than a decade directed a small but steady stream of treasury resources toward the development of its own GM crop varieties, as well as toward the simple backcrossing of foreign-developed GM traits into local germplasm. Between 1989-97, DBT spent a total of nearly 270 million rupees from the treasury (roughly $6 million) on plant and molecular biology research, with projects focused primarily on development of transgenic plants (Ghosh 1999). India's public research investment policies toward GM crops thus deserve to be classified here as "promotional." Tangible scientific payoffs from these investments have been slow to develop, however, reflecting both the modest total size of the investment plus some longstanding limitations within India's public sector research establishment. Without greater public spending on research, and without significant institutional reforms and policy adjustments, including some to help public researchers partner more effectively with the domestic and international private sector, India's goal of developing its own commercially useful GM crop technologies could remain elusive.

As noted earlier, India has a long and distinguished history of making effective public investments in agricultural research through the national institutes and universities that operate under ICAR. Public research by state level governments in India has also grown since the mid-1970s, even to rival total ICAR spending at one point in the mid 1980s. India's public sector investments in agricultural research, together with its formidable extension system, have had large impacts on productivity growth. By some estimates these impacts from research and extension have been larger even than those from government investment in rural infrastructure, education, irrigation, or power (Fan, Hazell, and Thorat 1999). Agricultural research and extension investments have been the source of roughly three fourths of all agricultural productivity growth in India in recent decades (Evenson, Pray, and Rosegrant 1999). Not all agricultural research in India has been public sector-led. After the mid-1980s, India's more liberal international investment policies helped attract more private agricultural R&D spending into the country. Between 1987 and 1995, private seed company R&D expenditures in the country increased from 41.7 million rupees to 154.9 million rupees. Still, total public sector agricultural R&D in India continued to exceed private sector R&D by a wide margin of roughly 6 to 1 (Pray et al 1991 and Kelley 1998).

Modern biotechnology was first introduced into the Indian research community in the 1970s, when a significant number of Indian scientists trained in the United States and Europe returned to take up work in India (Balasubramanian 2000). Significant government investments began quite early as well, with establishment in 1982 of a National Biotechnology Board. In 1986 this board was established formally as India's Department of Biotechnology inside the Ministry of Science and Technology (Sharma 1999). A significant part of DBT's mission has always been to distribute and finance specific research contracts, and in the area of agriculture DBT set up six centers for plant molecular biology in different regions of the country and has provided research funding to those centers- primarily through ICAR - for the improvement of specific priority crops such as rice, mustard, chickpea, pigeonpea, and wheat (DBT 1999).

DBT must secure its budget every year from the Planning Commission and the Ministry of Finance, and the resources it receives are quite modest, despite the fact that senior political leaders frequently list biotechnology as among the keys to India's future economic growth and prosperity. In 1998-99, the total research budget of DBT across all agricultural and non-agricultural areas was 1,040 million rupees, or roughly US$26 million. About 15 percent of this total (US$3.8 million) went for all plant biotechnology, including medicinal and aromatic plants, tree and woody species, and pilot plants for tissue culture. DBT's investments in *transgenic* plant biotechnology in 1998-99 totaled only about US$ 1.3 million.

With these modest treasury investments, DBT has been attempting to fund a significant spread of GM crop research activities, including not only basic transformation work but also research in extremely challenging areas such as abiotic stress (e.g., drought tolerance) and improved nutritional quality (e.g., protein quality, oil quantity and quality, or higher starch content). For example, ICAR is making efforts at Jawaharlal Nehru University (JNU) in New Delhi to engineer potato and rice with improved protein content, and elsewhere it is using GM to modify the starch content of potato and the Vitamin E activity of mustard and groundnut (Paroda 1999). A number of more standard GM applications are also being researched, including Bt varieties of rice, cotton, pigeonpea, potato, and mustard. National researchers have also worked with transformed eggplant, cabbage, cauliflower, and tomato.

As ambitious as this GM research agenda is, most of the work so far is at an early stage, or confined to laboratory or greenhouse trials only (Ghosh 1999, Table 1). Of all the ICAR GM varieties under development, DBT as of 2000 had approved field trials only for GM eggplant and mustard. The Indian Agricultural Research Institute (IARI) and the Central Institute for Cotton Research (CICR) have produced their own transformed Bt cotton, but as of 2000 just 100 plants had been grown in the laboratory only, and several years of backcrossing with local varieties would have to take place before any full scale field trials.

Over time, there is little doubt that India's national researchers will be able to develop and field test their own GM crop varieties suitable and ready for farmers to use. Prior to commercial release, however, some significant IPR impediments are likely to be encountered. Many of the transgenes India's researchers have been inserting are owned by foreign companies and have been made available for research purposes only. In the case of India's GM rice, the transgenes come from IRRI free of charge, but IRRI in turn has them for research purposes only via MTAs from the private patent holders. When the time comes to consider a commercial release in India, negotiations will be necessary with these international patent holders. ICAR presently has no clear strategy to break through such looming IPR bottlenecks, and it has yet to undertake the kind of IPR literacy program for its researchers that Embrapa earlier undertook in Brazil.

In several other respects ICAR may face difficulties pursuing its independent course toward GM crop technology development. ICAR continues to be dominated by conventional plant breeders, many of whom were trained years ago, well before the latest advances in GM technologies (Murthyunjaya and Ranjitha 1998). Younger scientists have been trained to do GM work for ICAR through the Rockefeller Foundation's rice biotechnology program, but often these people are hired away by the private sector, which pays more than twice the public sector salary. Younger scientists with good ideas can get support from DBT via ICAR, but excessive paperwork and long delays are routine. DBT's budget is not only small; it also tends to be spread too thin since each separate institute in the national system lays its claim to at least some financing. For example, India's rice transformation budget is divided up between at least five different locations within the national system, which weakens the impact at any one location.

India's national system takes pride in its independence yet some of its best work in the area of agribiotechnology has resulted from international donor support. Beginning in 1988 the Rockefeller Foundation became an important catalyst, helping India's national institutes and universities to build infrastructure, train younger scientists, yet this important source of external support has recently been phased out. The United States Agency for International Development (USAID) is also providing assistance, since December 2000 through a new joint research project run by Michigan State University and India's Tata Energy Research Institute (TERI) in Delhi, working to enhance the beta-carotene content of mustard oil to fight blindness caused by Vitamin A deficiencies. Also, the Ministry of Foreign Affairs of the Netherlands has funded an Andhra Pradesh-Netherlands Biotechnology Programme for Dryland Agriculture since 1996, engaging several ICAR institutes in a fruitful partnership to bring biotechnology applications to non-irrigated low-resource farmers producing sorghum, castor, groundnut, and pigeon pea.

This project, supported at 18 crores Rs over its first six years, has been more successful than the parallel biotechnology programme funded by the Dutch in Kenya, and discussed earlier in Kenya, so it is now scheduled to be extended for another 10-15 years. The programme focuses on low-end village level biotechnology applications, such as biopesticides and biofertilizers, but it has also given some financial support to ICAR researchers working on insecticidal GM varieties of sorghum and castor (IPE 2000). ICAR also has benefited from a substantial National Agricultural Technology Project (NATP) loan from the World Bank to upgrade its scientific skills, infrastructure and management procedures, and to facilitate better links between the component parts of the fragmented system, but little of this money seems to be reaching the labs of younger researchers in the biotechnology area. Meanwhile, links between ICAR and the private sector remain difficult to establish (Hall, et. al 1998).

Partnerships with the international private sector should be an attractive option in India, where national researchers could offer international companies not only their rapidly improving transformation skills and their long established conventional breeding skills, but also access to Indian germplasm from the 150,000 accessions and samples in India's National Gene Bank, established by

ICAR as a part of India's National Bureau of Plant Genetic Resources. When private companies show interest in working with ICAR or DBT, the response is muted however. The Monsanto Company at one point offered to share its Bt technologies with ICAR free of charge for transformation and production of orphan crops such as chickpea and pigeon pea within India, but ICAR showed little interest. And just as ICAR is leery of partnering with the private sector, most international companies are leery of depending too heavily upon ICAR, which has a reputation for slow movement of scientific results from the lab to the market place. The international private sector has shown its preference to work in India not with ICAR but with local joint venture partners in the private sector. In 2000 when Monsanto agreed to share its GM technology for high beta-carotene mustard with Indian scientists, it agreed to work through the private Tata Energy Research Institute, not through ICAR. Private seed companies in India have not only been ahead of ICAR in commercializing improved varieties such as hybrid rice; they are also known to have control over the highest quality hybrid germplasm. The international private sector will also be shy of working with ICAR on GM crops so long as the other policies of the Government of India - especially in the areas of IPRs, biosafety, and trade - remain highly precautionary or preventive rather than permissive. This is a loss, since the chances of getting a new GM crop variety through India's recently cautious biosafety screening process would probably be higher if the application for deregulation came via ICAR, rather than from a purely private applicant such as Mahyco/Monsanto.

Summarizing, the desire of the Government of India to promote GM crop technologies primarily through public sector research investments is in some respects logical and commendable, since private sector companies are unlikely to make the GM crop research investments needed to enhance nutritional traits for the malnourished, or to address abiotic stress production problems encountered in semi-arid regions by low resource farmers. Yet the success of this strategy will remain in doubt so long as the treasury resources being invested are so small, so long as the public sector institutions in question retain their current limitations, and so long as the policies of the government in other critical venues - biosafety in particular- remain preventive or highly precautionary.

## INDIA'S GM CROP OPPORTUNITY

The Indian people are far better fed on average than in the past, but 2.7 million children still die in India every year, 60 percent of them from diseases linked to malnutrition (Sharma 1999). A leading cause of malnutrition in India is poverty, and in rural areas a leading cause of poverty is low productivity in agriculture. Of the nation's one billion people, two thirds still gain their livelihood from farming, and seventy five percent of India's farmers are disadvantaged, with one hectare of land or less (Swaminathan 1999). Meanwhile continued population growth continues to increase the burden on Indian agriculture year after year.

The productivity of Indian agriculture improved in the 1990s compared to the 1980s, as the annual rate of growth of value added in farming increased from 3.1 percent to 3.8 percent, or roughly twice the nation's rate of growth of population (World Bank 2000). Poorly managed government marketing policies occasionally generated embarrassing public stocks of food, including 27 million tons of surplus wheat at the end of 2000. Yet the impression of national food abundance was misleading. During the decade of the 1990s, total food grain production in India did not increase at all on a per capita basis, and 230 million Indians remained food-insecure due to persistent poverty, linked most often to the low productivity of their agricultural resources.

Solving India's poverty and hunger problems will require more than just a boost in farm productivity, of course. A number of other issues will also have to be addressed, including persistent rural illiteracy, social marginalization, landlessness, and caste or gender discrimination. Even where low farm productivity is the problem, GM crop technologies might not be the solution. On India's drylands, where farm productivity is low in part due to poor soil fertility or scant rainfall, the GM technologies currently in use provide few new options to address such problems.

India's poorest farmers are those that live in dryland areas with less than 750 mm of rainfall a year who lack an ability to irrigate their crops. Non-irrigated farming in India still accounts for 67 percent of total cultivated area, and supports 40 percent of the population, plus two thirds of the nation's livestock. Average grain yields on non-irrigated land in India are only 0.7 to 0.8 tons per hectare, which is only one third the yield level on irrigated land (Singh and Venkateswarlu 1999). With yields on irrigated land now plateauing, India has little choice but to seek new technical solutions for its low-production farmers in dry, rainfed areas. In India 80 percent of millets and pulse crops are grown under rainfed conditions, and roughly 50 percent of rice is rainfed.

GM crops might seem an unlikely solution for farmers in hot drought-prone regions, since it has been far easier so far to engineer crops for specific resistances to pests or disease than to engineer the multi-gene traits needed to provide greater resistance to drought or heat. Yet India's producers of dryland crops (such as sorghum, groundnut, or pigeon pea) also face severe pest and disease problems, along with problems such as drought or heat. For groundnuts and pigeonpea, crop losses to biotic stress are actually greater than losses to abiotic stress (ICRISAT 1992). Pigeonpea farmers can sometimes lose their entire crop through damage from a single insect. Pod borers attack all pulses, and viral diseases are a widespread blight on India's dryland crops. Small dryland cotton farmers in India are devastated by bollworm infestations. Together with conventional breeding programs and improved training in integrated pest management (IPM), genetic engineering might help provide solutions to these biotic stress problems facing India's poorest farmers.

Environmental protection imperatives also argue for a GM crop revolution in India. The current practices of India's poor dryland crop farmers are damaging to rural ecosystems. If GM crops could produce yield gains for these farmers, there would be less need to clear new lands in rural India, plow fragile slopes, or destroy still more habitat. If farmers had insecticidal GM crops they also might escape having to risk their own health, pollute the environment, and kill so many non-target species as they do now with conventional chemical sprays. Farm chemical use is also a rural economic welfare issue in India, where cotton farmers currently spend 16 billion rupees annually on insecticide sprays. Vegetable producers in India currently suffer a $2.5 billion loss annually to insect damage, even while spending (on tomatoes, for example) $100-$200 per hectare on insecticides (Padmanabhan 2000)

Somewhat farther into the technological future, GM crops could eventually help address some of India's severe nutritional problems as well. Roughly 50,000 children in India go blind every year from vitamin A deficiency, while iron deficiency is a major threat to the health of women. The possibility of engineering iron-rich rice or vitamin A-rich rapeseed oil would become interesting in this context. Some traditional food crops in India that currently contain dangerous substances (neurotoxin in kesar dal, cyanide in tapioca, alfatoxins in groundnut) might also be rendered safer if genetic engineering could be used to "silence" these undesirable traits (Prakash 1999). And in a nation where refrigeration facilities are not yet abundant it might at some point be attractive to engineer fruits and vegetables less prone to spoilage. Tomato producers in India today lose 20-30 percent of their crop through postharvest spoilage.

Political leaders as well as scientists and technocrats in India have noticed these opportunities, and they now routinely endorse the potential contributions that biotechnology - including transgenic crops - might make to agricultural productivity growth and poverty reduction in the years ahead. In 1999, Dr. R. K. Pachauri, Director of India's Tata Energy Research Institute, stated that India's future increases in food production "would necessarily have to come from the application of biotechnology" (Pachauri 1999, p. 10). At India's 87th Science Congress in January 2000, Prime Minister Atal Bihari Vajpayee singled out information technology and biotechnology, plus "other knowledge-based sectors" as the propellers that would move India's economy ahead in the new century. At this same meeting Dr. R. A. Mashelkar, Director General of India's Council of Scientific and Industrial Research (CSIR) specifically endorsed biotechnology in agriculture as a means to turn farmers into high-productivity "knowledge workers" (BGE 2000).

While many top leaders in India have endorsed the value of agribiotechnology in general, and while scarce treasury resources have even been allocated to promote GM crop research within India's national agricultural research system, India's policies toward GM crops have hardly been promotional across the board, or even permissive. It was the original intent of biotechnology policy leaders in India to pursue an essentially permissive approach toward GM crops, yet this intent has recently been frustrated. Critics of GM crops were able to work within India's open and democratic political system to push for a precautious or even a preventive approach toward GM crops instead, especially in the area of biosafety policy. Indian biosafety authorities, somewhat like their counterparts in Brazil, ran up against forceful public criticism when they attempted to pursue a permissive approach toward the testing and release of GM crops. As of 2000 this meant that farmers in India, identical to their counterparts in Brazil and Kenya, had not yet been given official permission to plant any GM crops.

While GM crop critics in India have focused most heavily on possible biosafety risks, they have also attempted to block the technology by voicing strong objections to what they describe as the disadvantageous IPR implications of GM crops. To the extent that GM crops come with stronger IPR conditions attached, it is not surprising to hear voices of objection raised in India, a nation without a strong IPR policy tradition, least of all in the area of animals and plants. And to the extent that private multinational seed companies are the holders of so many of these IPRs, India's misgivings are that much easier to understand, given the culture of mistrust toward international companies that has long pervaded both India's government and its civil society. India's IPR policies toward GM crops, in part for these larger social and cultural reasons, have thus been extremely weak. As of 2000, India's Parliament still had not enacted even a basic plant variety protection law.

For decades India could get along without a strong IPR policy in the area of farm crops because it relied so exclusively on its own public sector scientists and government extension agents, rather than private investors or international private companies, to stimulate innovation in the area of agriculture. During the 1960s and 1970s, agricultural R&D in India was almost exclusively the business of the Indian Council of Agricultural Research (ICAR), a vast network of government institutes, directorates, research centers, and public universities. The private sector was effectively excluded through tight controls over germplasm exchange, technology imports, seed trade regulations, and restrictions on investment (Selvarajan, Joshi, and O'Toole 1999). In this environment, innovation was stimulated through public expenditure, rather than through the offering of IPRs to plant breeders. India operated without a plant variety protection law, and its 1970 Patent Act specifically excluded patenting of life forms such as plants (Mishra 1999).

This system served India well enough in the early years of the green revolution, and in some respects it continues to work well, as marginal economic rates of return to agricultural research in the state-dominated Indian system have remained high (Evenson, Pray, and Rosegrant 1999). By its own count during the 1992-96 Eighth Plan period, ICAR developed and released more than 2300 high yielding varieties and hybrids, including 452 new varieties of field crops ready for general cultivation. Yet by the 1980s, overall state support for agricultural research in India had begun faltering, and scientists in India began looking for ways to seek greater rewards through the private sector. Government attitudes toward the private sector also became less hostile. Starting in the early 1980s, the Government decided to provide its own breeder seed to private companies for multiplication and duplication. Foreign investment rules were relaxed, and in 1988 a New Seed Policy liberalized industrial licensing policies. As a result multinational seed companies began making significant investments, through India's own private sector companies. From the mid 1980s to the mid 1990s, 24 companies representing roughly 17 percent of all private seed concerns in India initiated technical or financial collaboration with foreign companies.

These liberalizations brought pressures in India to enact a basic plant variety protection law. Officials in the agricultural research establishment had concluded by 1991 it was necessary and prudent to move India's IPR policies closer to international standards. By 1993 a draft plant variety protection act (PVPA) was ready for submission to Parliament, one which measured up to the UPOV 1978 standard by creating relatively weak PBRs and by preserving the farmer's privilege to replant and exchange saved seed. The drafting of this basic PVPA preceded final negotiation of the TRIPS agreement in WTO, but when India formally ratified TRIPS in 1994 these issues became linked, since India now had an international obligation to have basic IPR protections in place for plant varieties by January 2000.

This Government decision to move toward a conventional plant variety protection law in the context of TRIPS triggered a surprisingly emotional debate in India's Parliament. The first draft of the PVPA was criticized by the private seed industry for being too weak, but NGOs claiming to represent farmers groups warned it was far too strong and would lead to a private expropriation of farmers' rights. Revised drafts were produced in 1996-97 in order to address the farmers rights issue, and Cabinet approval for a revised draft was secured in October 1997, but under still more NGO criticism Parliament continued to stall, and still more redrafting was initiated. The version of the PVPA that was working its way through Parliament in 2000 was a version produced in December 1999.

These PVPA debates in India would have been contentious even if GM crop technologies had never been invented. In India, the experience of colonization was an especially bitter one and sensitivities toward foreign ownership of any part of the national economy - especially by foreign companies - remain high. Foreign corporate ownership of genetic resources in India - GM or otherwise - is an especially emotional issue. Advocates of the PVPA have argued that IPR guarantees for plant breeders will stimulate useful new innovations in India, bring in foreign capital and technology, and at the same time provide for improved protection for the 150,000 accessions and samples of crop germplasm in the hands of India's own National Gene Bank. Existing plant varieties that are already in the "public domain" will also not be affected by the law. Yet critics highlight what they describe as the danger of a foreign corporate takeover of India's own national seed industry, or the private use of IPR options to gain controlling ownership of the plant genetic resources on which India's farmers depend. An Indian NGO calling itself the Gene Campaign argues prominently that foreign multinational companies would be able to appropriate and exploit India's genetic material without adequate compensation if western-style IPRs in this area were ever to be permitted (Sahai 1999).

Fueling these fears after 1991 was the increased presence of international companies staking ownership claims within India's food economy. Between 1987 and 1995, even though a PVPA had not yet been passed by Parliament, the share of private seed sales in India made by firms with foreign ownership increased from 10 percent up to 33 percent (Pray et al 1991; Pray and Kelley 1998). Several IPR actions taken outside of India were also represented as threats to national genetic sovereignty. In 1992 the W.R. Grace company secured a patent in the United States for a distinctive chemical formulation of a naturally occurring pesticide from neem trees. Anti-corporate activists in India began asserting that international companies such as Grace were planning to use such patent protections to appropriate for themselves the locally developed folk-knowledge that had long been available to India's farmers and rural communities for free. Grace argued that this charge was unfounded, since the patent was on a new chemical formulation implying that it did not prevent traditional farmers all over the world from continuing to use neem extract as they always had. Nonetheless the neem case enflamed popular anxieties regarding the foreign appropriation of local knowledge, and led to an international struggle by environmental NGOs to remove the patent. In May 2000, the European Patent Office did revoke Grace's 1995 patent in Europe for the neem oil extraction process, but litigation against the patent in the United States was sure to prove more difficult.[1]

Some of India's most respected leaders in the area of agricultural research have shared the concern that patent protections for plants, or even a national move toward a conventional PBR system, might leave the nation's poor farmers at a disadvantage. Rural communities in India have for thousands of years employed their own on-farm seed selection practices to breed a highly diverse stock of plant varieties nicely attuned to local conditions. Under a PBR system, why should IPR protection go only to the professional breeders (working either within international companies or national institutes) who routinely use these already improved local varieties as the basis for their breeding programs? Arguing that indigenous knowledge systems are similar to general scientific information, in that they are a valuable part of public knowledge, Dr. M.S. Swaminathan, India's most acclaimed agricultural scientist and the first winner of the World Food Prize, has helped popularize the notion that communities of farmers are as entitled to IPR protections for their efforts as professional breeders. Largely in response to his leadership, and as early as 1989, the FAO Conference in Rome adopted his concept of Farmers Rights.

Swaminathan also saw to it that India's own draft PVPA law would incorporate a farmers' rights provision, in the form of a institutional mechanism designed to ensure the rights of farming communities to financial compensation for any earlier contributions they might have made to the plant varieties that will finally be protected under the act. The 1999 draft of India's PVPA (in Chapters X and XII) thus certifies the rights of villages and local communities to claim contributions to the evolution of plant varieties, and to then receive appropriate financial compensations paid from a centrally managed National Gene Fund (Government of India 1999). In anticipation of this provision coming into effect, the Government of India began compiling a formal register of the geographic origins of all plant genetic resources that have been or might be used by professional breeders. Registering this indigenous germplasm also puts it in the public domain, thus preempting private bioprospectors from subsequently using the PVPA to take commercial advantage of this germplasm without compensating local communities.

As innovative and ambitious as these provisions for financial compensation to farmers might be, they have not gone far enough to satisfy India's most determined critics of UPOV-style plant variety protection. In 1998, India's activist Research Foundation for Science, Technology and Ecology (RFSTE) headed by Vandana Shiva, who is an internationally known opponent of GM crops, the green revolution, and all other market-led scientific advances in Indian agriculture. RFSTE proposed an alternative Biodiversity Rights and Protection Bill that would give local farming communities not just financial compensation but actual property rights over the crop varieties to which their traditional on-farm breeding efforts had made contributions. The Bill specifies that local communities would share these property rights with the Central Government of India so that private companies suspected of seeking to pirate away valuable genetic materials would not be able to approach local communities separately to gain access to these materials on unequal terms (Cullet 1999). The intent of this proposal was to reject any movement in India toward a western-style IPR system in the area of plant genetic resources; when the Government persisted with its PVPA approach, RFSTE initiated a public interest litigation action (PIL) against the Government.

The issue of GM crops only adds new complication to this already intense internal debate in India over plant variety IPRs. In 1998, when the Monsanto Company of the United States purchased a 26 percent share of India's own Maharashtra Hybrid Seeds Company Limited (Mahyco), a Monsanto executive was quoted in the Indian press saying "We propose to penetrate the Indian agriculture sector in a big way. Mahyco is a good vehicle." (*The Economic Times*, New Delhi, April 26, 1998). Opponents of transnational corporations within India took this as a direct challenge, and began directing harsh criticism at all of Monsanto's GM crop technologies, especially the "terminator gene" patent it had recently acquired, which was presented as a direct threat to the tradition of seed saving in India. It was Monsanto's misfortune that word of this new gene use restriction technology (GURT) reached India just at the time limited trials of Mahyco's Bt cotton were being authorized for the first time. A Canadian based NGO, the Rural Advancement Foundation International, or RAFI, spread the alarm that Monsanto's terminator gene might be turned loose in India. Local NGOs opposed to international companies and non-traditional farm technologies were thus mobilized easily against GM crops. In November 1998, a local political leader in Karnataka state who had previously gained attention by attacking both the Cargill seed company and a Kentucky Fried Chicken restaurant in Bangalore, staged for the media a brief attack on some of Mahyco's Bt cotton field trials. Soon thereafter local communist party leaders in Andra Pradesh pressured the Chief Minister into calling for an end to the Mahyco Bt cotton field trials underway there as well.

Gene use restrictions technologies had not yet been inserted into GM crops anywhere let alone the Bt cotton being tried out in India, but this issue made Monsanto and Mahyco easy targets for NGO and opposition party criticism. The mere existence of a terminator gene patent seemed to confirm suspicions that international seed companies were seeking to take away from India's farmers their traditional right to replicate seed on their own farms. In India, 92 percent of all wheat seed planted is home grown, and 88 percent of paddy rice seed.

NGOs and globalization critics feared that India's small farmers would be pressured by Monsanto or Mahyco into purchasing expensive GURT seeds, only to discover too late that they had to keep purchasing them year after year. The same argument could have been made against conventional non-GM hybrid seeds, which are currently used and repurchased annually by small as well as large farmers in India growing maize, sorghum (jowar), millet (bajra), sunflower, cotton, and vegetables, but the terminator technology was far more inviting as a proxy for criticizing profit-making foreign companies. Monsanto announced in the fall of 1999 that it was not going to commercialize the technology, yet by then the damage was already done, including damage to the Government's continuing efforts to move the PVPA law through Parliament.

Enactment of the PVPA in India was also made more difficult by links to the TRIPs agreement in WTO. Many of the same groups who saw IPRs for plants as the opening wedge for external corporate domination of India also feared the WTO, which is often criticized in India as an instrument used by rich countries to bully poor countries. Under the TRIPS agreement India was technically obliged to change its patenting system from processes to products, and to have a plant variety protection law in place by January 2000, but this deadline came and went without parliamentary action.

Prospects for eventual Parliamentary passage of some version - probably a weakened version - of the Government's PVPA bill were nonetheless strong in 2000. The National Democratic Alliance government sent its most recent version of the bill to Parliament for approval in December 1999, just before its January 2000 WTO deadline for TRIPs compliance. It was then referred to a thirty-member joint committee of both Houses which redrafted the bill so as to strengthen its farmers' rights provisions, by recommending that appeals be heard not by the High Courts but by a specially formed tribunal. The delays continued, but WTO was unlikely to criticize India for not having a PVPA in place so long as the delay seemed to reflect the slow movement of democratic procedures.

This failure of India through 2000 to set in place any formal IPR protections for plants was not, however, the principal reason that GM crop technologies were not yet in use by farmers. Because of the size of India's commercial seed market, private international companies have at times been eager to bring GM technologies into the country even in the absence of IPR guarantees. Their preferred means for doing this has been to seek to introduce hybrid varieties, which carry their own inherent biological protections against seed saving and replanting.

It has not so far been a corporate unwillingness to bring valuable proprietary GM crops into India that has kept this technology out of farmers' hands. India's weak IPR policies do, however, place some limits on what international private companies are willing to do, as well as on what India's own breeders have an incentive to do. Because of weak IPRs, private companies do not like to conduct advanced research in India labs. In its Bt cotton venture the Monsanto company conducted no plant transformations within India itself; instead it brought in hybrid cotton seeds that had already been transformed abroad. Nor did not begin field trials until it owned a secure share of a local corporate partner that had been entrusted to do the backcrossing. A Belgian biotechnology firm, Plant Genetic Systems, was almost as cautious when it decided to bring its transgenic "seedlink" technology into India, to develop an improved *brassica* (mustard). It invested an initial US$1 million in a joint venture with a trustworthy local private sector partner, Pro Agro, and even so it limited its transfer of transgenic materials to local Indian hybrids only.

If India wishes to partner with foreign companies to secure access to GM technologies for plants other than hybrids (for example, glyphosate resistant soybeans), passage of a credible plant variety protection law will be critical. Up to a point, bilateral contracts can be used to compensate for a weak internal

IPR law, yet without at least a UPOV 1978 plant breeders rights system in place, India's public sector researchers risk being passed over as unsafe technology development partners by the more dynamic international private sector. Guarding India's own nationally managed germplasm accessions from private appropriation could also become more difficult, public registration strategies notwithstanding. For these reasons India's Council for Scientific and Industrial Research (CSIR) has been campaigning for a stronger IPR and patenting system in India, in agriculture as well as in other sectors.

In the longer run, IPR guarantees even stronger than those contained in the current draft PVPA may be necessary if India wishes to keep pace in GM science. The draft PVPA has been criticized by the private seed companies as unattractive for their purposes in several respects. The draft PVPA lays a basis (in Chapter X) for compulsory licensing after three years to meet "the reasonable requirements of the public for seeds." It creates a registration process open to delay and challenge which could add another three years to the already long time period (currently 6-7) years required in India to get a new seed variety completely de-regulated and eligible for commercial release. And by embracing the UPOV 1978 standard the PVPA does not provide scientists with sufficient incentive to innovate at the molecular level, since protection of GM varieties may not be extend to included essential derivations from those varieties. The President of India's private Seed Association, while a supporter of the PVPA, has stated his view that it "does not provide sufficient protection for bioengineered plants." (Selvarajan, Joshi, and O'Toole 1999, p. v). Private seed companies would of course prefer product patents as the best means to encourage innovations in biotechnology, but Indian scientists working at the molecular level inside the national research system also want stronger standards; they would prefer the UPOV 1991 standard over the weaker UPOV 1978 standard contained in the current draft PVPA .

Without a plant variety protection law formally in place, India as of 2000 must be classified as taking a preventive stance toward GM crops in the IPR area, and even after the draft PVPA is eventually enacted India's posture in this area will be no more than precautionary. Still, it has been India's biosafety policies, not its IPR policies, that have most conspicuously slowed down the nation's GM crop revolution.

## BIOSAFETY POLICY

Biological safety has not been, historically, a strong public policy thrust in Indian agriculture. Rural biological systems have long been threatened by low resource farmers who clear new lands and destroy natural habitat in order to graze animals or plant crops. Mismanagement of irrigation water has left once fertile croplands poisoned with salt. Careless application of chemical nutrients has contaminated rivers, streams and ponds, making water unsafe for human consumption and poisoning fish and amphibian species. Improper use of insecticides has killed non-target species while making target pests increasingly resistant to the poisons. Even after the enactment of India's powerful 1986 Environmental Protection Act (EPA), such conventional biohazards from farming were seldom put under effective regulation. It is thus somewhat surprising to see Indian officials paying so much attention to the biological safety of GM crops. India's official interest in genetic resource IPRs would have been strong even without GM crops as an issue, but the nation's current official interest in rural biological safety is to a large extent a product of the GM crop revolution itself.

India's first embrace of a formal national biosafety policy dates from the first emergence of rDNA agricultural and human health technologies in the 1980s. We might expect, given India's strong development aspirations and its prior history of low attention to the rural biohazards, that the nation's biosafety policies toward GM crops would have been permissive or even promotional, rather than precautionary or preventive. Instead, the Indian Government adopted from the start a set of procedures that ensured significant precaution, and as implemented - in the context of a growing global controversy over GM crops - these procedures sufficed to slow the progress of the technology considerably. As of 2000 farmers in India were not yet permitted to grow any GM crops, because none had yet received final approval from the nation's biosafety regulators.

The Indian Government first issued rules and procedures for handling GM organisms in December 1989. The Department of Biotechnology, inside the Ministry of Science and Technology, published these rules and procedures in January 1990 (Department of Biotechnology 1990). These "Recombinant DNA Safety Guidelines" were revised slightly and republished as "Revised Guidelines for Safety in Biotechnology" in 1994 (DBT 1994). They describe the biosafety measures that must be undertaken in India both for contained research activities and also for large scale open environmental release of genetically altered agricultural and pharmaceutical materials. A subsequent 1998 revision of the Guidelines elaborated procedures for screening transgenic plants and seeds for toxicity and allergenicity (DBT 1998).

India's crop biosafety Guidelines were written to require a screening of GM crop technologies for scientifically demonstrated risks, borrowing heavily from the demonstrated risk approach employed by USDA's Animal and Plant

Health Inspection Service (APHIS) in the United States. This has given the Guidelines considerable permissive content on paper. Yet the implementation of the Guidelines is ultimately shared with India's Ministry of Environment, a procedure which inclines India's policy somewhat more toward precaution. The Guidelines create two separate review committees: a Review Committee on Genetic Manipulation (RCGM) which is empowered to approve (or not approve) applications for all *small scale research activities* in India designed to generate information on transgenic organisms, and a Genetic Engineering Approval Committee (GEAC) empowered to approve (or not) *large scale research activities, plus actual industrial use or environmental release* of all GM organisms.

Because the RCGM mostly screens research activities for biosafety, it is composed of representatives from India's leading public universities and research Institutes, including the Indian Council of Agricultural Research (ICAR), the Indian Council of Medical Research (ICMR), the Council of Scientific & Industrial Research (CSIR), and others. It is constituted by the Department of Biotechnology (DBT), and its member-secretary is adviser to DBT. It has thirteen members in all, and operates by majority vote. In contrast the GEAC is a statutory body under the Ministry of Environment and Forests (MoEF) and is empowered to approve or disapprove all large-scale use and environmental release of GM organisms. GEAC is thus India's most powerful biosafety policy gatekeeper. It is chaired by the Additional Secretary of MoEF, co-chaired by an expert nominee from DBT, and it includes representatives from DBT, from the Ministry of Industrial Development, the Ministry of Science and Technology, and the Department of Ocean Development. GEAC is designed to perform more than a technical function. It is the lead interministerial body empowered to shape - by consensus - the Government's final disposition toward large-scale use and environmental release of GM organisms. The GEAC is empowered to authorize or prohibit, conditionally or unconditionally, the import, export, transport, manufacture, processing, use, or sale of any GM organism (Ghosh and Ramanaiah 2000).

This Indian biosafety system was intended to be permissive at the research stage, where RCGM was given the final word. Since RCGM is constituted by DBT - the agency explicitly tasked with promoting biotechnology in India - its membership comes from the very research institutes (in some cases the very individuals) who conduct biotech research and who rely on DBT for funding. When RCGM hands the regulatory process over to GEAC, after the research stage, a precautionary bias takes over, since GEAC is chaired by MoEF. This built-in tension between permissive research versus precautionary release might have some advantages in theory. By allowing research to go forward, India's process is capable of generating the necessary empirical data (e.g., from field trials) needed to make sound final release decisions(Ghosh 1997). India's system may also be potentially useful for avoiding last minute jurisdictional disputes between biotechnology commissions and environmental ministries, of the kind seen in Brazil between CTNBio and IBAMA. In India, biosafety regulation stays legally under the EPA and within the jurisdiction of GEAC from start to finish. The public sector dominated Indian system also excludes representatives of private industry from direct participation, another contrast to Brazil, and it requires that RCGM send a Monitoring Committee to visit the experimental sites that are being employed by applicants to generate biosafety data (DBT 1998).

India's Guidelines are quite thorough regarding the biosafety questions that must be answered prior to commercial release of GM technologies. In the area of transgenic plants, basic information must be provided on the characteristics of the donor organisms, the vectors used, the transgenic inserts, and then on the resulting transgenic plants themselves. Through lab, growth chamber, greenhouse and field trials, evaluations must be generated by applicants regarding toxicity and pathogenicity, possibility and extent of transgenic pollen escape and transfer to wild relatives, and consequences for both the environment and human and animal health. The data may be generated in India or anywhere else (Ghosh 1997). Guidance is provided on proper (biosafe) procedures for conducting these experiments, along with the proper format for presenting this information as a formal registration document. Some of the data required by India's procedures go well beyond biosafety. RCGM is required in the latest version of the Guidelines to solicit from applicants and assemble information "on the comparative agronomic advantages of the transgenic plants" to assure GEAC that the technologies under review are "economically viable" as well as environmentally safe (Department of Biotechnology 1998, p. 6). While this might seem an added barrier to approval, it could also favor approvals in some cases by allowing evidence of social benefits to be presented alongside evidence of possible environmental risks.

This being the structure of India's biosafety review system for GM organisms, how has it operated in practice? In the area of health care, a number of GM products have been reviewed successfully with little controversy. As of March 2000, a total of ten different transgenic health care products in India (two locally produced, 8 imported) had moved all the way through the system and had been cleared for biosafety deregulation by GEAC. Social confidence regarding the biosafety of GM health care products has been consistently high in India. In the area of transgenic plants, however, the system has not moving so smoothly. Under criticism from NGOs, even RCGM has been forced into a highly cautious posture, and as of 2000 GEAC had yet to grant a commercial release for any GM crops.

India's biosafety review system for GM plants was pushed into this cautious posture in 1998, when RCGM was accused by NGOs in India of having exceeded its mandate by giving approval to limited field trials for Mahyco/Monsanto's Bt cotton plants. The original Guidelines had been ambiguous as to where RCGM's authority to manage small-scale research activities would end and where GEAC's authority to manage large-scale activities and "environmental release" would begin. When Mahyco applied to conduct contained field trials of its Bt cotton in 1998, RCGM assumed it had the authority to approve such trials, in part because it had earlier approved contained field trials of transgenic mustard in 1995 without incident. Yet the Bt cotton field trials approved for 40 locations in nine different states India in 1998-99 soon became a focus of intense national controversy, as noted above. NGOs in India, led by RFSTE, claimed that only GEAC had authority to approve field trials of GM crops, since even contained and limited trials constituted an environmental release. In February 1999, RFSTE then filed a PIL against the DBT for the manner in which it had authorized the trials (and also against Mahyco plus several other ministries), asking - without success - that further field trials be blocked.

RFSTE's challenge to the Bt cotton field trials was motivated not so much by any specific biosafety worry. The "Bollgard" Bt cotton seeds that Monsanto had brought in to cross with local Indian varieties had previously undergone extensive testing and monitoring in the United States to meet the biosafety standards set by APHIS and EPA. Bt cotton had been grown successfully in the United States on a commercial scale since 1996, and by 1998 it covered 40 percent of total U.S. cotton area. By 1998 it was also being grown widely and successfully in Australia, South Africa, and China. Bt cotton was potentially attractive in India as a solution to the bollworm infestations that farmers could no longer control with chemical sprays due to resistance in the pest population. This emergence of a real pest resistance problem from non-GM technologies made it harder to invoke hypothetical pest resistance problems as a reason not to plant GM cotton. resistant pest population in India in response to non-GM cotton production techniques made it harder to invoke hypothetical "superbugs" as a reason not to use GM cotton. Food safety risks were also hard to invoke for Bt cotton, since cotton is an industrial crop rather than a food. RFSTE was using biosafety rules and procedures as its means to try to block Bt cotton, but the core of its argument against this GM crop was that it was being introduced by the Monsanto company from abroad, the company which owned rights to a dreaded "terminator" technology.

Yet RFSTE also sought to make its case by pointing to procedural irregularities in RCGM's approval of the field trials. RCGM was faulted for having given field trial authority directly to a private company applicant, and for then allowing that applicant to launch the trials directly on leased farmers' fields, rather than on government research farms under direct public sector management. While this would seem a logical way to handle an application filed by a private company, RCGM might have known it would be seen in India as trusting the private sector too much. Second, RCGM had not secured advance approval from government authorities in several of the states where the trials were located, and in at least one of those states the trials even began before a local biosafety committee had been fully constituted. These several irregularities were exploited by RFSTE, and left RCGM and DBT politically isolated when fears regarding the terminator gene subsequently broke. Nor did it help that RCGM's procedures in this case had been largely non-transparent. RCGM could point out that there was no formal obligation to publish deliberations or invite public comment prior to approving the limited trials, but when the trials became controversial RCGM and DBT were criticized all the more heavily for having operated behind closed doors (Raj 1999). In the midst of this controversy in 1999, DBT's biosafety Guidelines were amended to clarify that GEAC's approval would be required for all field trials larger than one acre per location, or larger than 20 acres per year nationally. DBT also directed that the 1999 field trials for Mahyco's Bt cotton would take place either on university research farms or under closer national institute supervision. This, plus a DBT-led public awareness campaign in 1999, plus ministerial promises that the terminator gene would not be allowed into India, helped calm the public mood considerably.

Although India's Bt cotton trials went forward under protest and criticism, they did manage to generate important evidence of the GM plant's effectiveness against bollworm infestations. These pest attacks on cotton in India had emerged as a visible social issue due to the growing resistance of the pests to chemicals, and the resulting desperation of some small farmers. Excessive spraying of diluted and adulterated chemical insecticides - in some parts of India as many as 14 sprayings a season - had brought on the pest resistance problem, and some poorer cotton farmers in India who had borrowed heavily at a high rate of interest to buy chemicals that no longer worked were falling heavily into debt. Their plight came to public attention when bad weather worsened the pest crisis in 1998, and as many as 500 farmers resorted to suicide as their only way out (Sharma 2000). The first 1998 field trials of Bt cotton in India had seemed to confirm, meantime, that Bt varieties could provide at least a short-term solution to the pest control problem. DBT reported in 1998 that on average (in controlled field trials planted in eight different states in India) Bt cotton was able to reduce insect damage dramatically: yields for Bt cotton were 40 percent higher than for the non-Bt controls, and with an average of five *fewer* chemical sprays (Ghosh and Ramanaiah 2000; James 2000). For an Indian farmer with 5 ha of cotton, such a reduction in chemical use from switching to Bt cotton might by itself represent a savings of roughly Rs 2000 per crop (about $50).

RCGM was pleased and reassured by the 1998 field trial data, but nonetheless asked Mahyco in 1999 for ten more field trials, seeking more data on commercial advantages to farmers and also on a hypothetical geneflow concern, the possibility that insects could take Bt cotton pollen far enough away from the plant to result in unwanted gene transfer to other plants. Only after these additional trials had been completed did RCGM, in April 2000, formally express its technical confidence in Bt cotton. RCGM noted that its latest evidence showed farmers growing Bt cotton obtained 25-75 percent higher yields, while using six fewer chemical sprays, and with no evidence of harm to crops in adjacent plots.

Having received this approval at the research stage from RCGM, Mahyco promptly applied to GEAC for permission to begin large scale trials, and in response to this application, GEAC did approve large scale field trials for Bt cotton (up to 85 total hectares) in July 2000. Additional acreage was also permitted for seed production, in anticipation of a possible commercial release as early as 2001. To appease protesters GEAC made large scale field trials conditional on obtaining an independent Indian laboratory certification that the cotton plants did not contain the "terminator" gene. Even so, RFSTE was antagonized and submitted an additional petition against the trials. GEAC's prompt field trial approval increases somewhat the likelihood of an eventual commercial release for Bt cotton in India, although the first release may be only within a tight acreage restriction, once again on grounds of biosafety caution.

Has the Indian system been precautionary in this case only because the Bt cotton is a Monsanto variety? A second application working its way through India's system, from another company (ProAgro-PGS) to develop transgenic hybrid mustard, also met a cautious response from RCGM. Contained field trials of GM mustard began in 15 different locations in 1995, followed by open field trials, and none of these field trials was ever disrupted by an anti-GM activist, yet the process of bioscreening was highly precautionary just the same.

In 1999, India's regulators asked ProAgro-PGS for one more year of field trials to screen for effects on soil micronutrients, an issue that had never been highlighted before. ProAgro-PGS hoped for eventual deregulation from GEAC, but recognized that it may be a conditional release only, perhaps with a requirement (hard to enforce) that farmers planting the GM mustard seeds separate themselves at least 40-50 meters from non-GM fields, as an extra precaution against unwanted geneflow. While India's anti-GM NGOs may be focused almost exclusively on keeping Monsanto's products out of the country, biosafety regulators in India have thus been highly precautionary toward all GM crop applications, whether from Monsanto or not.

## TRADE POLICY

Since GM crops are politically controversial in India, the government is under pressure from NGOs to impose a GM-free trade policy requirement on the nation's commodity imports and exports. So long as no GM crops are being planted within India on biosafety grounds, such an official GM-free trade policy can be relatively easy for the government to embrace and implement. A policy of standing aloof from international GM commodity trade is also relatively easy to embrace in India because the nation has, for decades, tended to stand somewhat aloof from all international food commodity trade in its official pursuit of "national food self-sufficiency." Ever since India's bad experience with excessive dependence on international food aid in the 1960s, political leaders have sought to avoid not only renewed concessional dependence on world markets, but commercial dependence as well.

India's policy aversion to international food trade of all kinds is reflected in the fact that the nation has recently accounted for roughly 10 percent of total world agricultural production, but less than 1 percent of world commodity trade (Sharma 2000). With continued income growth in India, demand for imported grains is likely to grow, yet actual imports will continue to be slowed by significant trade barriers at the border. As one example, India recently imposed an 80 percent duty on rice to curb the influx of what it called "cheap grain."

India occasionally imports small quantities of corn, but it strictly controls these imports with a tariff rate quota imposing a 60 percent duty on imports larger than the quota. In the case of wheat, India allows imports only in rare cases to offset specific internal transport cost problems, for example to allow some less expensive imported wheat to reach some coastal flour mills in the southern part of the country.

India also exports very little wheat, despite its occasionally large internal surplus stocks. In 2000, when India's excess wheat stocks reached 27 million tons, efforts were finally made to clear the stocks through export, yet these were frustrated in part by the known presence of a wheat crop disease - "Karnal Bunt" fungus - in some parts of India, which has put wheat from India on the import ban list of some thirty other countries (APBN 2000a).

Given these larger aversions in India to free international food commodity trade, and given the absence so far of any GM crop production within India itself, imposing an effective ban on the import and export of GM commodities has been easy enough for the Indian government to manage. Under India's official GM Guidelines it is GEAC which must approve any large scale import of GM commodities, and so long as GEAC has not approved the commercial planting of GM crops at home it may naturally wish to go slow on approval of commercial imports from abroad.

Politicians in India will resist GM crop imports as well. Ever since the 1998 controversy over field trials for Monsanto's Bt cotton, government ministers have routinely defended themselves with statements that GM foods are not being grown *or imported* into India, and will not be imported without proper safeguards.

India's official GM crop import ban was tested under unusual circumstances in the summer of 1998, when the nation experienced a highly publicized food safety emergency unrelated to GM crops. More than 50 people in Delhi died after eating a contaminated locally grown (and non-GM) mustard oil, the most popular cooking oil in northern India.

The government acted by banning sales of mustard oil pending safer packaging. To make up the shortage, it had to authorize temporary imports of soybeans and lower its import duties on soybean oil (*Wall Street Journal, December 8, 1998, p. A10).*

Anti-trade and anti-GM NGO leaders responded to this action with a letter to the Prime Minister, alerting him to the fact that some of the soybeans about to be imported could be GM. The government sought to finesse the issue on this occasion by arranging to import soybeans that were already split and hence suited only for oil production, and not planting. Because the soybeans were not intended for environmental release and since they were not explicitly labeled GM, the Government on this occasion managed to avoid seeking formal import approval from GEAC. In the future it may be more difficult for the Government of India to avoid engaging GEAC in approval of bulk commodity imports.

The January 2000 Biosafety Protocol obliges exporters (under Article 18) when shipping LMOs internationally to label those shipments as "may contain LMOs" and "not intended for intentional introduction into the environment." Identifications of this kind are more likely to trigger a mandatory review in India by GEAC, or at least more vigorous complaints by NGOs if GEAC is again avoided. NGOs in India opposed to GM crops and foods can take extreme positions on the import question.

In June 2000, RFSTE reacted with outrage when it learned that some of the corn-soya blend food aid that was being imported via CARE and CRS to provide relief to Indian victims of a super cyclone in the Eastern coastal state of Orissa, had come from the United States and was thus likely to be "genetically contaminated." RFSTE castigated the United States for "using the Orissa victims as guinea pigs for GE products" and called on the government of India to explore alternative sources of food aid (RFSTE 2000).

India's policy regarding imports of GM germplasm for research purposes has so far been permissive rather than preventive. Some extra steps are required when importing GM materials for research, and some bureaucratic delays are encountered, but the imports themselves have never been held back.

This partly reflects the fact that RCGM, rather than GEAC, has final authority to clear transgenic imports for research purposes (DBT 1998, p. 8). Technically, any importer of GM germplasm must also get phyto-sanitary clearance from India's relevant quarantine authority, the National Bureau of Plant Genetic Resources (NBPGR), yet NBPGR typically acts on the recommendation of RCGM and conducts a phyto-sanitary evaluation, as a routine precaution against import of diseased or infested materials.[5] RCGM is a committee dominated by representatives from the government institutes doing some of the research, so its permissive stance on imports is not surprising, yet private companies as well have had little trouble bringing GM materials into the country if the purpose is research only.

The import decision process in India is prone to a jurisdictional dispute between RCGM and GEAC, similar to that discussed earlier in the case of field trials. RCGM is authorized to approve small scale imports only, and GEAC alone can approve large scale imports, but there is not yet a clear definition of the dividing line between the two. Tiny quantities of 10 kg or less have until now fallen clearly under RCGM's jurisdiction, but a potential for conflict over larger shipments remains. India's Guidelines document empowers GEAC to give approval or disapproval (from an environmental vantage point) of all large scale "import, export, transport, manufacture, process, [or] selling of any microorganisms or genetically engineered substances or cells including food stuffs..." (DBT 1990 p. 17) The encompassing language of this provision again suggests how little the Government of India has been willing to entrust this new technology to private sector market forces alone, let alone international market forces.

In export markets, India has on occasion faced a temptation to use its GM-free status to seek price premiums. India is a small exporter of soybean meal (1.5-2.2 million tons per year, in recent years), and it has recently promoted these meal exports (also sunflower meal and rapeseed meal) as "GM free" for sale to overseas markets, such as Japan, Indonesia, Thailand, the Philippines, the Gulf countries, and the Middle East (APBN 2000b).

Yet most of these sales are for animal feed purposes, so price premiums have been difficult to secure. Indian meal exporters have nonetheless begun hoping that Asian countries such as Thailand, because they export chickens to the GM-conscious European market, will soon see the advantage of taking feed imports from a GM-free supplier such as India, rather than from nations that plant GM crops such as the United States.

India can thus be classified as embracing a trade policy toward GM crops which is fully "preventive" by the definition in use here: a de facto blockage on GM commodity imports, coupled with occasional efforts to use the nation's GM free status to seek premiums in export markets. The strength of this policy will be tested if and when GM crops are finally given biosafety approval for planting by Indian farmers, particularly food crops (not just cotton). At this point the nation's overall GM free status will disappear, so costly internal market segregation and labeling procedures will be required in order to offer GM free guarantees to foreign customers. Avoiding these costly trade-linked procedures could become an additional reason for India's GEAC to go slow on the final approval of internationally traded GM food crops for planting by India's farmers.

## CONCLUSION

We can see from this summary a seeming contradiction between India's public research policy toward GM crops, which is promotional, versus its IPR, biosafety, and trade policies, which are precautionary or even preventive. Is there an explanation for this apparently inconsistent mix of policies?

Given India's unmet farm productivity needs, and given the endorsement of GM crops by its own research community, why the extreme precaution in these other areas? India's sluggishness in developing a plant variety protection policy is perhaps understandable because it links to so many issues other than GM crops, and because it has not in any case been the highest barrier to getting GM crops into the country.

But how can we explain India's strong precaution toward GM crops in the two areas of biosafety and trade policy?

A comparison to some of India's policies at the outset of the green revolution in the middle years of the 1960s can perhaps shed some light here. In 1965, when the Center for Improvement of Maize and Wheat (CIMMYT) in Mexico and the International Rice Research Institute (IRRI) in the Philippines first made high yielding varieties of wheat and rice available to India, a political debate ensued within the country not so different from the current debate over GM crops. Indian groups on the political left and those claiming to speak for the interests of small farmers were highly skeptical then toward the green revolution HYVs, just as these same groups today tend to be highly skeptical toward GM crops.

The new Mexican wheat varieties had not yet been widely tested in India itself, so critics argued it might be dangerous to begin planting them on a large scale without more information regarding their local performance and impact. The new varieties also called for more sophisticated management practices that small farmers might not be able to master, and a use of purchased inputs small farmers might not be able to afford, so fears arose that only big commercial farmers would gain, and India's rural poor might be made poorer still. The new varieties also required more fertilizer than India itself could produce at the time, so critics feared an implied dependence on fertilizer sales made by private international agrochemical companies located in Europe or the United States.

Despite these many understandable and sincerely voiced concerns, the Government of India set caution aside in 1965. Under courageous political leadership from Agriculture Minister C. Subramaniam and others, India decided in 1965-66 to import 200 tons of Mexican wheat and appropriate quantities of high yielding rice to launch an ambitious seed production and farm demonstration campaign, with a goal of planting 13 million hectares to the new varieties within five years. Subramaniam was told by western experts and by India's own Planning Commission that his goal was too ambitious, and that he should not try to move so fast (Subramaniam 1979, pp. 45-51). Subramaniam's plan to move ahead quickly with new green revolution seeds was ultimately supported by Prime Minister Indira Gandhi and then aggressively implemented, despite continued internal opposition.

His approach won the policy debate on this occasion because the alternative to taking the new seeds was continued dependence on imported food aid from the United States, which had always come with unwelcome diplomatic strings attached. As Subramaniam recalls telling his critics at the time, "Would you like to continue this dependence on America for imported foodgrains? Was that preferable to raising domestic production? Instead of importing food was it not better to import fertilizer and plant protection chemicals to help raise production?" (Subramaniam 1979, p. 28).

When GM crop technologies first became commercially available in the 1990s, India was no longer significantly dependent on food imports from abroad, either as food aid or even as commercial purchases. The success of the green revolution itself had eliminated the need for such imports. GM crop advocates have therefore not been able to use Subramaniam's trump argument: that an embrace of GM crops is necessary to free India from food import dependence.

Instead of importing wheat as food aid, India's government today is burdened by surplus stocks of wheat.

Several hundred million Indian villagers are still poor and poorly fed because they cannot improve the productivity of their own farmlands or protect their crops from pests and disease, and this could well be seen as a legitimate reason to move ahead quickly with GM crops. But this line of argument is hard to develop since most of India's leading advocates for the poor within the NGO community are opposed to GM crops on other, just as many of them were earlier opposed on similar grounds to the green revolution HYVs.

RFSTE's Vandana Shiva, the leading NGO critic of GM crops in India today, had earlier made an international career out of criticizing high yielding non-GM crops in India (Shiva 1991). It was perhaps to be expected given these political dynamics, that a far more cautious approach to the new technology would prevail. In the current case the caution is expressing itself in India through controversies over the biosafety of GM crops, but the same larger background debate about the wisdom of relying on the international private sector or of importing new agricultural technologies from the West is alive and influential just below the surface.

But what is the opinion of India's farmers toward GM crops? Until farmers in India are given permission by the government to grown these crops, this important question will be impossible to answer. Farmers in India developed a strong positive opinion toward high yielding seed varieties in the 1960s and 1970s once they were given access to the new seeds, and once India's own farmers voted in favor of high yielding varieties by adopting the new seeds to quickly and so widely, the policy debate continued among some NGOs and non-farmers but otherwise became largely irrelevant.

Today in India various self-styled farm leaders have taken pro-GM or anti-GM positions in the policy debate, but until GM crops are released for commercial use by actual farm communities the views of farmers will not be known. By slowing the movement of GM crops into farmers' fields, the government is postponing the day when the nation's most important stakeholders in the transgenic crop revolution will be able to develop and express their own informed opinion.

# 8

# Genetic Resources and Improvement of Crops

Arid zone of northwestern India is bestowed with a number of diverse plant species including food crops, shrubs, grasses, trees and horticultural crops. These species possess a unique adaptation to prevailing climatic conditions of the regions as they have been cultivated here over centuries under harsh conditions leading to development of specific germplasm resources/ecotypes adapted to this zone.

These resources consist of landraces, cultivars, genetic stocks, elite germplasm and wild and weedy relatives of cultivated plants. They play a very critical role in providing food, fodder and nutritional security in arid regions. The major arable crops are pearl millet, legumes including mung bean, moth bean, clusterbean, pigeonpea, sesame and Indian mustard. Seed spices chiefly include coriander, cumin, fennel and fenugreek and main horticultural crops are *ber*, date palm, pomegranate, *aonla*, *bael* and *gonda*.

The main grass species are *Cenchrus ciliaris*, *C. setigerus* and *Lasiurus sindicus*. The trees species being improved are *Prosopis cineraria*, *Tecomella undulata*, *Acacia senegal*, *Salvodora oleoides* and *S. persica*. Many research institute located in arid zone have been enduring to address several issues toward genetic improvement and management of genetic resources of these species. We present here salient findings in germplasm management and genetic improvement of these species. The issues that are pertinent in future research are also addressed.

## GENETIC IMPROVEMENT OF TRESS AND SHRUBS

Trees and shrubs play a great role in arid ecosystem as both human and livestock depend on these especially in drought years. The productivity of the deserts can be increased by using genetically improved stocks of trees and shrubs. Identification of mating system in arid zone trees and shrubs: Information about the breeding system of trees and shrubs is usually lacking. Studies showed that *P. cineraria, T. undulata, S. oleoidis, S. persica, Acacia senegal* and *Azadiracta indica* have both systems of pollination. Self -pollination is dominant in most of the species, except *T. undulata* and *A. senegal.* Reproductive capacity in these trees was found to be very low (1-2% even less than that). Flowers appearing in February - April in A. indica fall before fertilization while those appearing late in April and May set into fruits. *In S. oleoides* and *S. persica* flowers appearing in December-February fall while those appearing in March-April set into fruits. Similarly, *T. undulata* flowers of November-January fall without forming fruits Seed setting during theses periods was only when the environment was conducive for the natural pollination in terms of air temperature, relative humidity and number of pollinators.

Mostly trees and shrubs of this area are cross pollinated, but when the environment is not favorable for the pollinators and the dehiscence of the anthers, the same species produce seed with self pollination, but with very low in number. So, in case of trees where pods and seeds are consumed, the late flowering genotypes should be selected for better fruit set. Exploration, collection and ex-situ conservation of diversity: Large area of Rajasthan including districts of Jodhpur, B1armer, Jalore, Bikaner, Churu and Jaisalmer and Gujarat were surveyed for distribution of species like Prosopis cineraria, Tecomella undulata, Acacia senegal, Salvodora oleoides, *S. persica, Lawsonia inermis, Cassia angustifolia* and *Commiphera wightii.* From different sites, plus trees/shrubs were identified and their seeds/cuttings were collected plant wise. Progenies of these plants were evaluated in nursery. Poor ones were rejected and the plants of better progenies were transplanted in the field as half- sib plant progenies trials in case of plants using seeds, and clones for clonally propagated species like mehndi and guggal to study the variation, heritability, genetic advance of different growth parameters over different years. More than 4000 plants of different accessions of different multipurpose trees and shrubs are being evaluated at CAZRI. Clonal seed orchard of *P. cineraria* has been established.

Selection methodology for trees of arid zone: In *T. undulata,* estimates of family heritability, single- tree habitability and genetic gains for family and mass selection indicated that the gains were low with both the methods for tree height. Components of variance tend to stabilize after fourth year of growth. So, in this tree species, mass selection will not be effective and the selection of the plants should be done after fourth year. In *P. cineraria* and *S. oleoides* high values of heritability due to families and single trees coupled with high genetic gain suggested these trees can be improved by simple mass selection which is perhaps the easiest breeding method in forestry.

To maximize gains there should be control of pollen parents, which can be achieved by progeny trials, and selecting the best families and best plants within the families. Similarly, combined selection for effective improvement in *A. senegal* has also been advocated. Seed orchards: The half-sib progeny trials can be converted into seed orchards by removing the poor families and poor trees within the family. In *P. cineraria* eleven half-sib progenies trial establishment in 1984 by CAZRI was converted into seed orchard. Before roguing the number of surviving trees varied from 11 to 31 out of 42 trees planted per progeny. About one third of the trees which had poor growth, crooked stem, infested with galls were removed. Total of 134 trees were retained in the seed orchard. Height, collar diameter and DBH before and after roguing of the trees in the orchard showed that performance of the retained trees was 17.6% better for tree height, 19.6% for collar diameter and 25.3% for DBH than the performance before rouging.

## GENETIC IMPROVEMENT OF HORTICULTURAL PLANTS

The Indian arid zone is characterized by high temperature and low and variable precipitation which limit the scope for high crop productivity. However, these conditions greatly favour the development of high quality in a number of fruits such as date palm, *ber*, aonla, bael, pomegranate, kinnow, lasoda. It is now realized that there is a limited scope for quantum jump in fruit and vegetable production in the traditional production areas. The recent awareness regarding the potential of these ecologically fragile lands for production of quality produce has not only opened up scope for providing sustainability for the people of this region, but also for bringing in new areas to increase horticultural production.

The area expansion and yield potential of arid horticultural crops has increased because of availability of new varieties and advancement in production techniques in arid region. Constraints in arid horticulture: The soils of arid region are very poor in fertility. The soils often have high salinity. The ground water resource is not only limited owing to poor surface and sub-surface drainage, but is also saline in quality. The irrigation water resources in the region are seasonal rivers and rivulets, surface wells and some runoff water storage devices (e.g., *nadi, tanka, khadins*) and canal irrigation in arid region.

Thus, the water resources in arid region are limited and can irrigate hardly 4% of the area. The annual average rainfall in the Indian arid regions is very low and varies from 100 mm in north-western sector of Jaisalmer to 450 mm in the eastern boundary or arid zone of Rajasthan. In some parts of arid region, occurrence of frost is also common features during winter season which affects vegetative growth of plants as well as productivity, quality of fruits especially in ber and aonla. There is no heat tolerant variety of arid horticultural crops which should be developed to achieve higher production. One of the major bottlenecks in development of horticulture scenario in arid parts of the country is lack of sufficient quality seed and planting material. The weak processing infrastructure and marketing of horticultural produce, lack of proper cold storage facility and knowledge and equipment for grading and packaging of fruits and vegetables are major constraints in production.

Genetic resource management: Diversity of major arid horticultural crops is being conserved in field repository at CIAH, Bikaner and CAZRI, Jodhpur. Among 318 *ber* (*Ziziphus mauritiana*) genotypes, the *ber* varieties Gola, Seb, Umran, Kaithali and Banarasi Kadaka are performing well under hot arid climate. Varietal evaluation resulted in delineation of varieties for arid zone under different situations Early maturing varieties of *ber* such as Gola and CAZRI Gola were identified for rainfed areas, where as cultivars such as Umran, Seb, Illaichi, Tikadi, Banarasi, Dandan, etc. were found suitable under partially irrigated situations. Tikadi, a local cultivar of *ber* was found to be resistant to fruit fly (Singh and Vashishtha, 1984), but its fruit size was small. The cultivars Seb was found to be a good mid-season cultivars, but it was susceptible to fruit fly. The F1 of Seb X Tikadi was found to be resistant to fruit fly but the dominance of Tikadi prevailed as regards to fruit size was concerned. Therefore, it was back-crossed with Seb, which resulted in acceptable fruit quality along with fruit fly resistance. Similarly hybridization was done between Seb and Katha to improve the shelf-life of Seb. The F1 of Seb x Katha had 3 days more shelf life than Seb. Studies were also conducted on biosystematics of 59 cultivars of *ber* (Vashishtha and Pareek, 1993) which helped to divide these varieties into three groups, viz. erect, semi-erect and spreading type based on morphological characters. The cultivars Umran, Ajmeri and Katha was found to be isogenic, but named differently at various locations.

Out of 154 genotypes of pomegranate (*Punica granatum*), Jalore Seedless, Ganesh, G-137, Mridula, Phule Arakta are the better genotypes for yield and quality. Evaluation of 15 cultivars resulted in identification of the cultivars better adapted in arid regions with limited irrigation facility. Under breeding programmes, 13 cross combination were tried. The F1 of Ganesh x Jalore Seedless, Bedana-1 x Ganesh, Bedana II x Jalore Seedless, Jalore Seedless x

Ganesh and Ganesh x Bedana Bosec were found promising. The varietal evaluation of aonla (*Emblica officinalis*) revealed that the NA-7 (Neelam) is a prolific bearer followed by Chakaiya and NA-6 (Amrit). Among 16 bael (*Aegle marmelos*) genotypes, NB-5 and NB-9 performed well under irrigated hot arid ecosystem. A five-year-old budded plant of NB-5 produced about 40 fruits tree-1 while NB-9 produced 29 fruits tree-1. The fruit size of NB-5 is smaller (1 kg fruit-1) than NB-9 (1.4 kg fruit-1). The fruit quality is excellent in both the varieties. New varieties were introduced during 1983-84. Evaluation of cultivars revealed better performance in Laljeet Sambhipuri, Dhara Road and Faizabadi local. Later three accessions were added in 2003, of which two were wild type having small fruit size, but with strong flavour highly desirable quality for processing.

In National repository of date palm (*Phoenix dactylifera*), sixty indigenous and exotic cultivars/genotypes are maintained and evaluated for different horticultural traits. The cultivars Halawy, Khalas, Khuneizi, Barhee, Zahidi, Medjool, etc. are suitable for cultivation in hot arid region. In addition to this, a number of minor underutilized crops are also being conserved and evaluated for different traits (More and Singh, 2008). Out of 20 cultivars introduced and evaluated (Vashishtha, 1981), Halawy, Barhee and Muscat-2 have been found suitable for raw eating at doka stage while those of Khadrawy and Medjool for preparation of Chuhhara and Halawy, and Jahidi for pind khajoor. The work on germplasm survey, collection and evaluation in custard apple was started in 1985.The survey of Annona growing areas revealed that in Rajasthan it occurs in wild forms in Chittor Garh, Kubhal Garh, and Aravali hill ranges. Promising seedlings were identified from the germplasm collection of various locations. Chittorgarh Local was most promising followed by Bangalore local. Ten *gonda* accessions collected from different parts of Rajasthan and adjoining areas were evaluated for fruiting behaviour and fruit yield for seven years. There was not much variation for fruit shape, size and growth habit of the plants but a great deal of variability was invariably observed in total fruit yield per plant. Four promising types were identified based on total fruit yield. Accession No. G2012, G2021, G2025 and G2061 were found to be high yielding.

Six accessions and a variety (Pant Manohar) of Karonda have been introduced at CAZRI. They could be grouped into two broad categories based on fruit colour. The selection CZK2000-1 has pink peel on white back ground with 3.4 g mean fruit weight and ovoid in shape. Another selection CZK2001-17 has purplish green colour peel with mean fruit weight of 4.8 g and elongated ovoid in shape Survey of Khejri (*Prosopis cineraria*) germplasm for desirable pod types for culinary purpose resulted in identification of two superior types of plants having longer pods and thornless. Large fruited lime (*Citrus latifolia*) is also a is high yielding fruit species with mean fruit weight of 75 g with 56.5% juice and seedless but its foliage was found to be susceptible to hot winds during summer.

Twenty five accessions of Kair (*Capparis decidua*) were collected from differenced parts of Rajasthan. Recently a yellow flowered type, different from normal scarlet red flowered and a thornless type have been added.

Realizing the importance of vegetable crops, particularly under exploited and less popular, and having commercial production potentials in arid and semi-arid regions, systematic research work on germplasm collection, conservation and utilization was started in 1993 at CIAH, Bikaner. In this context, intensive crop specific surveys in target variability pockets and explorations were undertaken in arid and semi-arid regions and a large number of land races, semi-cultivated and popular types of mateera, kachri, snap melon, beans and some perennial horticultural species of vegetables potential were made over the years for systematic evaluation, characterization and conservation of indigenous germplasm. A number of variety of arid vegetables like mateera, kachri, snap melon, Indian bean and cluster bean has been released for cultivation.

## Fruit and Vegetable Improvement

CIAH has released 16 varieties of arid fruits and vegetable crops which includes, Thar Bhubharaj, Thar Sevika, Goma Kirti of ber, Thar Shobha of khejri, Goma Aishwarya of aonla, Thar Samridhi of bottle guard; AHW-19, AHW-65, Thar Manak of mateera; AHK-119, AHK-200 of Kachri, AHS-10, AHS-82 of snap melon, AHC-2, AHC-13 of kakri and Goma Manjri of clusterbean. A number of varieties of arid horticultural crops are at prerelease stage. These include AHRM-1 of round melon; AHG-13 in clusterbean, Indian bean AHDB-3, AHDB-16 and Sward bean (AHSB-1) besides some promising lines in ber, mulberry, lasora and ker, have been identified for evaluation and release (More *et al.*, 2008). Varietal variation in endurance to drought has also been observed in horticultural crops.

Early ripening cultivars seem to escape stress during receding soil moisture after monsoon. Ber cultivars Gola, Seb and Mundia for extremely dry areas, Banarasi Kadaka, Kaithli, Umran and Maharwali for dry regions, and Sanaur-2, Umran and Mehrun for comparatively humid regions have been recommended. Apart from morphological parameters, plants should also have physiological parameters for endurance to drought for commercial cultivation in this region. Some physiological parameters identified in *ber* are no mid day depression in photosynthetic rate, low rate of transpiration, maintenance of leaf water balance, growth, canopy development, dry matter allocation, high water-use efficiency, etc. It has been demonstrated that plant having capacity for drought endurance are able to maintain turgour, dry matter allocation, leaf and fruit growth even under low soil moisture level. Biotechnological tools for improvement: Micropropagation has been commerciallized in many ornamental crops and herbacious fruit crop species worldwide.

However, its wide spread commercial use is still limited in fruit tree species because of several inherent problems of *in vitro* culture system. Keeping in view these inherent limitaions of micropropagation of fruit tree species, a new concept of micropropagations of fruit tree species has developed using single or double node explants from the mature trees. Direct morphogenesis of shoot and root formation was achieved in lasoda (*Cordia myxa*), mulberry (*Morus alba*) and lime (*Citrus aurantifolia*) using single or double node explant having physiological active axillary buds. In another study with citrus, direct shoot and root formation was achieved in double node explant within 35 days of culture period. These results conform the production of plantlets within a short period completely eliminating subculturing process. Thus, using this technique of micropropagation, the fruit tree species can be multiplied in vitro with minimizing inherent problems of tissue culture in greater way to obtain a large number of genetically identical, physiologically uniform and developmentally normal plantlets preferably with high photosynthetic or phototrophic potential to survive the harsh ex-vitro condition. Attempts have been made for mass multiplication of ker through tissue culture technique since it is a hardy plant and suitable for hot arid environment. Work on date palm tissue culture is being done at various places in the country However; some good results have been achieved through organogenesis and embryogenesis in date palm tissue culture at CIAH, Bikaner.

## Prospects in Arid Horticulture

Presently, there is a scope in expansion of horticultural crops in arid region and it has vast potential for changing scenario of horticulture of the country. Vast land resource, surplus family labourers, increasing canal irrigated area, developing infrastructural facilities, plenty of solar and wind energy, etc. are the strengths in arid region for development in arid horticulture. Further, minimum pressure of diseases and insects in the region is good scope for production of seed and planting material. Ber is commercially grown in more than 80,000 ha. with production of 0.9 million tonnes under semi arid and arid regions. It requires more attention for value addition. Pomegranate production (area 0.132 millions ha) and production (1.14 million tonnes) is increasing very fast in dry part of the country. Since, it has tremendous scope of export of this crop from semi arid and arid regions of the country.

Presently, the export value from pomegranate is Rs.70 million. The crops like fig, custard apple are also coming very well under dry land conditions. At present, fig is cultivated in more than 1,000 ha area in Maharashtra, Karnataka. Likewise, custard apple is grown in about 50,000 ha area in the state of Maharashtra, A.P., Karnataka, Rajasthan and Tamil Nadu. Aonla is a medicinal fruit plant and cultivated in over 55,000 ha area producing 150 m t fruits/year. Date palm is most suitable fruit tree of hot arid region and it is grown in Rajasthan, Gujarat, Punjab and Haryana in over 12,493 ha producing 85,000 t fresh fruit. However, date is imported from Gulf countries due to its meager production. Bael is also an important fruit crop of semi-arid and arid regions. Now, attention is being given on its commercial production. India is the second largest producer of vegetables (125.88 million tonnes from 5.77 million hectare area in 2007-08). In case of vegetables in the same period the increase in area and production was observed to be 92.12 and 147.60%, respectively. In the reported period under major spices, the increase in area and production was recorded to be 6.98 and 83.73%. The per capita consumption of vegetable in the country has increased from 47 kg in 1984 to 76 kg in 2000, with annual growth rate to 2.9%. By the improvement in production technology in arid region, many seed spices likes Coriander, Cumin, Fenugreek, Ajowain, Fennel, Dill and Nigella, are being cultivated on large scale and also exported to earn foreign exchange. There is a vast potential of floriculture in some parts of Rajasthan because of low infection of disease and insects. The prospect of floriculture under hot arid condition is also important from seed and plant production point of view.

## GENETIC RESOURCES FROM ARID REGIONS

The arid zone, which forms a part of the 12 bio-geographical zones of India, is unique from the point of view of both floristic as well as climatic factors. The dry arid region in India encompasses the Thar Desert of Rajasthan and Kachchh and adjoining regions of Gujarat. The genetic diversity of this region is of great significance because of its adaptability to harsh environmental conditions. It is the storehouse of genes for stress tolerance and represents germplasm, which can be used as donor in developing hardy varieties. The diversity in this region is now becoming increasingly threatened in view of large scale over exploitation and overgrazing of the resources on the one hand; and change in consumption pattern, increasing irrigated agriculture, urbanization, mining, industrialization, pollution and climatic change on the other hand. Additionally, there is an inadequate knowledge of potential uses of this immense diversity for benefit of human. Therefore, there is an urgent need to conserve genetic resources of wild species and weedy forms which are of potential value for future need. Since 1976, the National Bureau of Plant Genetic Resources (NBPGR) is actively engaged in the conservation of germplasm of plants, their wild relatives and weedy forms from arid region.

## Germplasm Collection

In all, 17396 germplasm accessions comprising of cultivated crops and related wild species have been collected from arid and semi-arid regions of northwestern India. These include cereals (1191), millets and minor millets (1257), pulses and legumes (5942), oilseeds (2046), vegetables (1950), fruits (1507), spices and condiments (224), fiber crops (543), medicinal and aromatic plants (1095), forage and fodder (301), wild and weedy species (623), and other economically important plants (591).

## Germplasm Introduction

As on December 2008, 3688 germplasm accessions of different crops/ plants have been introduced in the arid and semi-arid region of northwestern India both from abroad and from other agro-climatic regions of India. Some of the notable exotic introductions include jojoba (*Simmondsia chinensis*), wild species of clusterbean (*Cyamopsis senegalensis, C. serrata*), salt bush (*Atriplex* spp.), 24 species of *Acacia*, while the introductions from within the country include budded ber, pomegranate, sesame, etc.

## Germplasm Conservation

A total of 729 germplasm accessions belonging to 107 species of vegetatively propagated and perennial crops are being maintained in the field gene bank at the Regional Station, Jodhpur. As on June 2009, 30560 germplasm accessions belonging cereals and millets (6709), legumes (13668), oilseeds (6005), economic plants (924), medicinal plants (572), horticultural plants (1650), fiber/fodder (120), released varieties (46), wild species (166) and other economically important species (165) are being conserved as active collections at the medium-term storage facility housed at the Regional Station, Jodhpur. This germplasm includes 10195 germplasm accessions from other research institutes of the region.

## Germplasm Registration

A genetic stock of clusterbean (CH/55-2) with foliaceous bracts in the inflorescence has been registered (INGR No. 08027). This germplasm accession produces foliaceous bracts, wherein the bracts are modified into small leafy structures. Besides, this genetic stocks for specific traits such as short internodes, determinate plant type, branched and all node cluster bearing, high gum content have been identified in clusterbean and an extra bold seeded accession in mung bean.

## Germplasm Distribution

One of the important steps in utilization of germplasm is the distribution of germplasm with specific traits for need based research to different crop based/region based institutes. As on December 2008, 17117 germplasm accessions of different agricultural crops (14029), horticultural plants (774) and other economic plants (2314) have been supplied for research purposes to different indenters throughout the country.

## Documentation

The publication of the germplasm characterization and evaluation are being made available to the breeders of different crops through the publication of catalogues, proceedings of the workshops.

## Future Perspectives

The challenges for the future in the area of plant genetic resources conservation are technical and scientific, socio-economical, legal and political, including public awareness. Therefore, continuous monitoring of biodiversity is essential for its conservation and sustainable utilization. The plant species in the desert ecosystem are unique, not only in their characteristic features enabling their survival and proliferation, but also in its utility value for human beings directly or indirectly. Thus survey, collection and conservation of useful and rare, threatened and endangered plant species are essential before they become extinct due to over exploitation. Conservation *in-situ* in protected habitats and *ex-situ* through seed gene banks and field gene banks, biodiversity gardens and cryopreservation should be the priority and NBPGR plays an important role in order to achieve this objective through survey, collection, characterization and conservation along with documentation and verification of the indigenous knowledge for its utilization.

## GENETIC IMPROVEMENT OF CROPS

### Pearl Millet

The arid zone of north-western India is a unique adaptation zone for pearl millet because of its inherent features like severe moisture stress, high temperatures and poor soil fertility. Arid zone represents approximately 25% of the total acreage of this crop in the country. At the national level, both the production and productivity have been increasing although the area has been declining. Average pearl millet productivity has increased from 408 kg ha-1 to 908 kg ha-1 during the last more than four decades. This allowed the local people to meet their pearl millet grain and stover needs from relatively smaller area releasing additional land for other industrial and cash crops like cotton, sunflower, maize and groundnut. The productivity gains have been achieved largely due to adoption of high yielding hybrids and cultivars. In contrast, hybrid adoption and productivity gains have been much more limited in the arid zone of the north-western India. Farmers in this zone still largely cultivate stressadapted traditional landraces on a large scale, mainly to minimize the risk of crop failure due to severe drought. The lack of adoption of currently available pearl millet hybrids is largely due to their inadequate adaptation to the marginal environmental conditions. Hence, increasing the productivity of pearl millet in the arid (A1) zone has remained a continuing challenge for breeders. Significant progress in pearl millet productivity in this region through crop improvement, however, is feasible. The likely approaches to be followed and the resources needed to address the various issues of genetic improvement of pearl millet for arid zone are presented here. Delineating the arid environments: Looking at the unique climatic conditions in arid regions of north-western India, the All India Coordinated Pearl Millet Improvement Project (AICPMIP) carved out a specific A1-zone within A-sub zone of north and north-western India in order to cater to the needs of arid regions. There are still large differences within arid regions in terms of prevalent environmental stresses or farming system needs, meaning large diversity in cultivar requirement within the arid zone. The variation in the amount and distribution of rainfall from east to west across arid region creates a wide range of soil moisture availability. In addition, as the environment becomes drier, the relative importance of millet grain stover becomes increasingly more important. Yadav *et al.* (2003) classified arid zone in three groups viz., extreme stress environments, moderate stress environments and near-optimum environments within arid zone according to production constraints and then described the breeding objectives for each sub-zone.

Role of adapted germplasm: A number of studies have established that the more marginal the conditions, the better was the relative performance of landrace-based materials compared to those bred for high yield potential under favorable conditions. Hence, AICPMIP has recognized the importance of adapted germplasm in breeding both hybrids and open-pollinated varieties. Hybrids HHB 67 and HHB 67-improved, which have been most popular hybrids in arid regions are based on the pollinators derived from landraces from western parts of India. Similarly, composites CZP 9802 and RCB 2, recommended for cultivation in arid regions, have been bred from adapted landraces from drought prone areas. A comprehensive list of pearl millet cultivars for A1 arid zone has been provided by Khairwal *et al.* (2009)

Need for genetic diversification: Landraces and elite populations have contrasting, but complementary combinations of traits. Landraces, possessing a fair amount of adaptation to abiotic stresses, often fail to capitalize on the additional resources and hence are inherently low in their grain productivity. In contrast, the elite materials with high grain yield potential might not necessarily possess adaptation to stress environments of arid zone.

There is enough evidence that crosses between landraces and elite exotic populations have certain advantages over their parental combinations and it is possible to combine adaptation with high yield potential through hybridization between adapted germplasm and elite genetic materials. The performance of populations bred by introgressing non-landrace germplasm into landrace backgrounds also increases the responsiveness of landrace germplasm to better-endowed arid zone environments (Presterl and Weltzien, 2003; Yadav, 2007; Yadav, 2008), but selection in such materials needs to be done in a typical arid zone environment to retain the adaptation of the landraces.

Indian national breeding programme has also given a high priority to diversify the hybrid parental lines in pearl millet specifically for arid regions. Currently, a large number of male-sterile lines (seed parents) and restorers are being utilized in hybrid development from a wide range of germplam sources with the result that several downy mildew resistant hybrids have been developed with high grain and stover yields with a wide range in maturity time.

Exploitation of heterosis in arid zone environments: It is generally argued that heterosis is best exploited under favourable environment. While this could largely be true, several studies conducted in arid zone have demonstrated that heterosis can also be effectively exploited to amalgamate drought tolerance and high productivity in order to improve both grain and stover yields in pearl millet under marginal conditions (Yadav *et al.*, 2000 & 2009) of arid zone. Average improvement in grain yield of landrace-based topcross hybrids over their landrace pollinators has often been in the range of 20% to 40%.

Requirement under arid conditions is that the improvement in grain yield should not be at the cost of stover production; total biomass productivity needs to be increased. Yadav *et al.* (2000) reported an average of 15% heterosis in growth rate in topcross hybrids based on landrace pollinators, which translated to a positive biomass heterosis where crop duration of the hybrid equaled that of the pollinator. The partitioning of this extra biomass to either grain or fodder appeared to be controlled by the harvest index of the seed parent, resulting in differential heterosis for either grain or stover yields, depending upon the seed parent used. Bidinger *et al.* (1994) and Yadav *et al.* (2000) suggested that it may be possible to meet farmers' needs for increased grain and stover production, while retaining critical adaptation to arid zone environments, by exploiting heterosis between locally adapted landraces and male-sterile seed parents that partition the extra dry matter to both grain and stover. The variation in biomass heterosis is the major determinant of both grain-yield heterosis and stover-yield heterosis (Bidinger and Yadav, 2009). The contribution of harvest-index heterosis to grain and stover yield heterosis has been observed to be of a compromising nature. Harvest index heterosis accounted for substantial variation in grain-yield heterosis, but positive heterosis in harvest index resulted in negative heterosis for stover yield suggesting that strategy for increasing grain yield by improving harvest index will not result in the desired outcome in the marginal arid environments where stover yield is also important. There are exploitable genetic differences among the male-sterile lines and landrace-based restorers for their ability to produce heterotic crosses for biomass (Bidinger *et al.*, 2003a; Bidinger and Yadav, 2009).

Opportunities for hybrids in the arid zone: The lack of adoption of current hybrids does not necessarily mean that farmers are averse to hybrids. Currently available hybrids don't meet requirements of grain and stover in drought years. This is because these hybrids were bred from the parental lines with little adaptation to the arid zone and/or to farmers' requirements in this zone. Early research assessed the performance of topcross hybrids made with conventional seed parents and adapted landrace pollinators in the arid zone. Results indicated that it is possible to achieve high yields in these environments, initially by producing topcross hybrids from elite male-sterile lines and landraces (Bidinger *et al.*, 1994; Yadav *et al.*, 2000), but obviously additional work was required to identify seed parents with acceptable adaptation, and to breed disease resistant restorers, with good general combining ability, from adapted landraces. Subsequently a large-scale assessment of all publicly available malesterile lines was undertaken to assess their combining ability for grain and stover yields, stability, responsiveness to improved environmental conditions, drought tolerance, downy mildew resistance and plant type of their hybrids (Bidinger *et al.*, 2002; Bidinger *et al.*, 2003b) to identify sets of male-sterile lines with characteristics necessary for arid environments, ranging from low-input subsistence environments to those with supplemental irrigation facilities (Yadav *et al.*, 2003b). The key trait for seed parents was a positive general combining ability for biomass (Bidinger *et al.*, 2003a) which can be combined with good adaptation from pollinators bred from landraces to increase both grain and stover yields and to maximize yield stability. Specific male-sterile lines possessing good general combining ability (GCA) for biomass have been identified.

A set of landrace-based populations adapted to arid zone, with a range of phenotypes and of diverse origin have been used as source material for breeding adapted inbred restorer lines. The lines derived from two landracebased populations showed differences in their combining abilities for adaptation and produced hybrids that comprehensively outyielded the best available checks under typical drought environments (Yadav *et al.*, 2009). The advantage with these adapted hybrids is that they have more chances of being delivered to farmers through well-established seed delivery channels of both private and public sectors in India. We have begun developing inbreds from a number of other restorer populations based on improved germplasm developed by AICPMIP programmes in the arid zone and the International Crops Research Institute for the Semi-Arid Tropics (ICRISAT). Role of molecular biology: Molecular marker-assisted breeding has been an important activity in breeding downy mildew resistant hybrids of pearl millet. The first product from this approach has been released for droughtprone environments in A1 zone as HHB 67-improved, which is a downy mildew resistant replacement for popular extra-early maturing hybrid HHB 67 (matures in 65 days). The male parent of this hybrid was improved by backcrossing quantitative trait loci (QTL) determining downy mildew resistance. HHB 67-improved marked the first marker-assisted product in public domain in India that was released and marketed. Because of a history of limited success in breeding for drought adaptation by conventional phenotypic breeding, molecular breeding has become a prime focus. Molecular mapping and genomics approaches offer new opportunities and strategies to dissect major genes and QTL underlying drought tolerance. Identification of genomic regions determining drought tolerance of arid zone adapted landraces is underway at CAZRI. Research at ICRISAT has identified QTL which had significant effects on pearl millet yield under drought stress (Yadav *et al.*, 2002; Yadav *et al.*, 2004; Bidinger *et al.*, 2005). Comparison of hybrids with and without these QTL showed that QTL-based hybrids were significantly, but modestly, higher yielding under terminal drought (Bidinger *et al.*, 2005).

## Arid Legumes

Arid legumes comprising clusterbean [*Cyamopsis tetragonoloba* (L.) Taub.], cowpea [*Vigna unguiculata* (L.)], moth bean [*Vigna aconitolifolia* (Jacq.) Marechal] and horsegram [*Macrotyloma uniformis* (Lam)] are predominantly grown in arid zone of Indian sub-continent. The arid legumes are important crops for food and nutritional security and livelihoods worldwide. These crops are known for versatile uses like food, fodder, forage, vegetable, medicine and soil health enrichment. Arid legumes are an important source of dietary proteins, essential amino acids and minerals for the predominantly vegetarian population in India. Being highly adapted to hostile agro-climatic conditions, the arid legumes exhibit higher degree of tolerance to biotic and abiotic stresses. Area, production and productivity of arid legumes in Rajasthan: The area of clusterbean in Rajasthan hovers around 25 lakh ha while the productivity is around 250 kg ha-1. The moth bean is cultivated over an area of about 14 lakh ha with productivity of less than 200 kg ha-1 while cowpea is being cultivated over an area of 1.2 lakh ha with productivity over 300 kg ha-1. Almost 80% acreage of clusterbean, 90% of moth falls in north western districts of Rajasthan viz., Ganganagar, Hanumangarh, Churu, Bikaner, Nagaur, Jodhpur, Barmer and Jaisalmer. The potential productivity of clusterbean is more than 2500 kg ha-1, of moth 700-800 kg ha- 1 and of cowpea over 1000 kg ha-1. Despite development of high yielding and early maturing genotypes of these crops during last 15 years production and productivity are still very low compared to their yield potential because of several factors:

1. Erratic and low rainfall.
2. Cultivation of these crops on marginal soils having low fertility.
3. Poor economic conditions of farmers.
4. Planting local genotypes having low productivity and long maturity due to inadequate supply/availability of the quality and certified seeds of improved genotypes.
5. High risk due to biotic and abiotic factors, large price fluctuations and no minimum support price also discourage the farmers to grow arid legumes.

6. Further, arid legumes are crop of secondary importance, if there are timely rains farmers prefer to grow pearl millet, sesame and mung on fertile and good soils instead of clusterbean, moth and cowpea.

Breeding for improved seed yield, plant types and earliness: Arid legumes have traditionally been cultivated under aberrant drought prone situations with least inputs and care. These crops have shown increased grain yield under existing situations with little fertilizer application with lesser agronomic inputs. For instance, in clusterbean recently developed varieties like, HG-884, RGC-1038, RGM-112, RGC- 1017, RGC-1002, RGC-1003, HGS- 365, HGS-563 and GRC-936 maturing in 90-100 days have yield potential of 1000-1500 kg ha-1 in rainfed conditions, while RGC-1031, RGC-986 and RGC- 197 (unbranched) are late in maturity (110-125 days) and suitable for high rainfall zones or irrigated and better soils conditions and have yield potential over 2.0 tones per hectares.

In cowpea, varieties developed in late nineties, Vamban-1, KBC-2, V- 585, V-130, V-240 and GC-3 are late maturing types (90- 100 days) with the yield potential of 900-1200 kg ha-1 while the varieties developed in recent years, viz., RC-101 (2001) and CoVu-702 (2005) are maturing in just 65-70 days with same yield levels. In moth, varieties maturing in 58-60 days (FMM-96), 60-65 days (RMO-40, RMO-225, RMO-257, RMO-225 and CAZRI Moth-3), 65-67 days (CAZRI Moth-2) and 70-75 days (CAZRI Moth-1, Jawala and IPCMO-880) days having erect, semi-erect and semi-spreading plant types possess yield potential of 600-1000 kg ha-1 compared with traditional types (Baleswar-12, Type-3 and Jadia) maturing in 90-100 days and yielding 200-400 kg ha-1 only.

Alteration in plant types concomitant with early maturity befitting the region and rainfall distribution is therefore desired. Research efforts in moth bean have resulted in the development of promising genotypes characterized with erect growth habit and early maturity (58-62 days) compared with traditional genotypes having spreading plant habit and late maturity (90-100 days). These genotypes, inspite of shortening the growth period by 30-35 days have great yield potential. For instance, mutant lines CZM32 and CZM18 maturing in 63 days may yield as high as 780 kg ha-1. These lines besides, improved plant types and being early, are a potent source of drought tolerance. CAZRI Moth-1, a variety developed from Jadia also represents semi-erect growth habit and source of yellow mosaic virus (YMV). Other genotypes of moth bean like RMO 40, RMO 257, RMO 225 and FMM 96 released for cultivation represent effective alteration in plant types and high yield. CAZRI Moth 2 and CAZRI Moth 3 released ion 2003 and 2005, respectively, have high drought tolerance and grain yield potential under extreme arid situations of arid western Rajasthan receiving very less rainfalls (200-300 mm).

In cowpea also, compact and erect plant types with earliness is required, as the existing varieties with bushy and trailing growth habit yield less. A growth period of 65-70 days is therefore, required. Some genotypes like C 480, C 653 and C 722 maturing in 70 days, and RC 101 maturing in just 65 days had been induced. Similarly, mutants from Pusa Falguni (V 37, V 38, V 585 and V 240) with compact growth habit have been developed. Some of the mutants developed by Bhabha Atomic Research Centre are also showing synchronous maturity with determinate growth habit, very high yield levels (1300-1500 kg ha-1) and maturing in just 60-65 days. The lines developed at the International Institute of Tropical Agriculture have been introduced in India and are performing well and may fit in wheat-rice cropping system of Indo-Gangetic Plains for summer cultivation as they are maturing in 60-65 days with grain yield of 1200-1500 kg ha-1. In clusterbean, HG 563 and RGC 936 maturing in 85-90 days with very good yield potential have been developed for irrigated and rainfed situations, respectively.

Further, development of unbranched genotypes like RGC 1066 and RGC 197 paved the way for clusterbean cultivation in irrigated areas having heavy soils as they are yielding over 3000 kg ha-1 and suitable for mechanized harvesting. These genotypes are also suitable for mixed and inter-cropping with irrigated (groundnut) as well as rainfed kharif crops (pearl millet and sesame).

## Issues and Future Strategies

The breeding strategies of these main crops of the arid regions need to address several issues which may include:

1. Breeding for need-based alteration in plant types, so as to make them more acceptable under existing unpredictable environmental fluctuations.
2. Curtailing the maturity period of these crops so as to match specifically with the regions, soil types, rainfall pattern and social needs of the inhabitants.
3. Breeding varieties with 20-30% higher biomass per unit water.
4. Development of plant types suited to aberrant situations of mid June to end of August plantings.
5. Development of varieties suitable for specific cropping systems: intercropping, mixed cropping, agro-forestry, agri-horti and agri-pasture systems.

6. Attempting more and more inter-specific crosses, particularly between cultivated and wild types to tap genes from wild types for adaptations and resistance to biotic and abiotic stresses.
7. Increasing clusterbean gum content from existing 31 to 35% with carried viscosity profile level through breeding strategy.

## Mung Bean

Mung bean (*Vigna radiata* L. Wilczek), also known as green gram, is an important short duration grain legume with wide adaptability is an important crop in arid and semi-arid regions. In the last 3 decades, mung bean has registered phenomenal increase both in area and production. The area has increased from 1.99 m ha in 1964-65 to 3.77 m ha in 2007-08 and it became third largest pulse crop in the country. The production has increased from 0.60 million tones to 1.56 million tones during the same period. This expansion in area has resulted due to development of relatively photo- and thermo-insensitive varieties which allowed a greater flexibility in their planting dates. Further the shortening of crop duration paved the way for its inclusion in a wide range of intensive cropping systems.

Production constraints: This crop is grown mostly during rainy season, however development of short duration and disease resistant varieties led its cultivation during spring/summer season in almost all parts of the country and during winter (rice fallows) in peninsular India. The major constraints in this crop are:

1. Insect pests during kharif/spring seasons causing major losses to the crop.
2. Mung bean yellow mosaic in kharif and spring in northern part and powdery mildew during winter season in southern part of the country.
3. Lack of high yielding varieties of efficient plant type for different situations.
4. Pre-harvest sprouting in rainy season which restricts its intercropping with other companion crops such as maize, sorghum, pigeonpea etc., grown at the onset of monsoon.
5. Lack of on-farm storage facilities and its vulnerability to store grain pests like bruchids. Resistance against the insect is not available in germplasm.
6. Low seed replacement rate (10%) of improved varieties.

## Varietal Development

The initial phase of varietal development had been aimed at improving locally adapted but genetically variable populations mainly by pure line and mass selections with major emphasis on traits other than yield. This resulted in the release of large number of pure lines, some of which are still cultivated in certain parts of the country. Establishment of All India Co-ordinated Pulses Improvement Project (AICPIP) in 1967 provided an access to pulse breeders to improved germplasm and to test improved breeding lines in multi-location evaluations across the country. As a result more than 110 varieties of mung bean were developed. During the last three decades, major emphasis has been to breed short duration, photo- and thermo-insensitive varieties coupled with resistance to biotic stresses viz., yellow mosaic virus and powdery mildew, which contributed significantly to the national production.

The first variety of mung bean T 1 has extensively been utilized in hybridization programme to develop mung bean varieties T 2, K 851 and T 44 and Sunaina. Being a short-duration and acceptable seed quality variety T 44 became very popular in spring/summer season. T 44 was used in hybridization programme to develop Pusa Baisakhi, Pant moong-2 and Pant moong-4. Pusa Baisakhi has been utilized to develop PIMS 4 and Jyoti. Two varieties of mung bean ML 1 and ML 5 were developed from PAU, Ludhiana, during the early 1970s and these were used to develop ML 131, ML 267, ML 337 and ML 23. During the same period a variety Mohini (S 8) was developed through hybridization between T 2 and BR 2. Iranian germplasm PS 16 and S-8 were used in hybridization programme to develop KM-1, which has become a very popular variety of Southern zone of the country. Large seeded varieties of mung bean, Pusa Vishal, Pant mung 5 and SML 668 were developed from the selection of AVRDC material. Mutation has also resulted in development of Pant mung 2, MUM 2, Co 4, LGG 407, LGG 405 and BM 4.

Three varieties of mung bean Pant mung 4, HUM 1 and IPM 99-125 were also developed through wide hybridization between mung bean and urdbean. Pedigree analysis of mung bean varieties revealed that very limited genetic variability has been exploited in mung bean breeding programme (Katiyar *et al.*, 2007). Variety T44 has been used extensively in mung bean varietal development programme in India and it is one of the parents for more than 16% mung bean varieties.

Impact of improved varieties at farmers' field: To demonstrate the production potential of newly released high yielding varieties, 1133 FLDs were conducted in different states. On an average, improved genotypes recorded grain yield of 781 kg ha-1, which was 22% higher than the local varieties.

Role of wide hybridization: The nature of research reported on wild germplasm related to grain legumes in comparison to cultivars and landraces has contributed to increased opinions of the value of wild germplasm. Studies have suggested that about 150 species of *Vigna* exist in nature and 22 *Vigna* species including cultivated and wild are found in Indian gene centre.

Various reports indicate that many desirable traits having great potential for mung bean improvement are available in wild or related species. However, most of these species remain unutilized due to complications such as crossing barriers and linkage drag. A wild accession of *Vigna radiata* var. *sublobata,* PLN 15, is found to be the potential donor for higher number of pods per plant and seeds per pod. Efforts are in progress at AVRDC to utilize *V. radiate* var. *sublobata* for resistance to bruchids (*Callosobruchus chinensis* and *C. maculates*) which is controlled by a single dominant gene. There are some successful examples where the wide hybridization led to the development of cultivars with added advantage for practical use. Five mung bean cultivars, HUM 1, Pant Mung 4, IPM 99-125, Pant mung 6 and IPM 02-3 have been developed from mung bean x urdbean crosses. Several other derivatives from mung bean x urdbean crosses exhibit many desirable features such as MYMV resistance, lodging resistance, synchrony in podding and non – shattering. A few mung bean x rice bean and mung bean x *V. radiate* var. *sublobata* having high degree of resistance to MYMV were also recovered.

***Priorities and Thrust Area:***

- Incorporation of photo- and thermo-insensitivity traits.
- Incorporation of resistance to pre-harvest sprouting and bruchids.
- Pyramiding of genes for resistance to major insect pests (Thrips,
- Jassids and pod borer) and diseases (YMV, PM and CLS) for which resistance is not high in cultivated germplasm.
- Identification of diverse germplasm source for important economic traits and plant types.
- Accessing desirable genes from wild species.
- Development of short duration varieties with distinct vegetative and reproductive phase for fitting the crop in narrow windows in cerealcereal cropping system.
- Improved seed production and supply systems (seed village, community seed bank, etc.).

## Development of New Plant Type

The presently available plant type in mung bean is largely photo- and thermo-sensitive, overlapping vegetative and reproductive growth habit with low harvest index and low grain yield. To remain a commercially competitive crop mung bean has to fit in high input cereal-cereal cropping systems. For this plant type that are determinate, photo- and thermo-insensitive, early maturing, high yielding (1500-2000 kg ha-1) with balanced vegetative growth, less plasticity in vegetative growth and high harvest index, and resistant to lodging is to be developed. A plant type with thick stem with two–three erect branches, short internodes, having small and thick leaves needs to be developed. Good seedling vigour, a clearcut distinction between vegetative and reproductive phases and high harvest index will be essential component of the plant type.

## Pigeonpea

Pigeonpea [*Cajanus cajan (L.)* Millsp.] owing to its constitutive traits like deep root system and immense flexibility in morphological/ phenological characters is inherently suitable for arid and semi-arid agro-climatic conditions. Pigeonpea with an annual production of around 3 m tones constitutes about 5% of the total world production of pulses. A major part of this is produced in India (90%). Traditionally the crop has been growing in India along the Indo-Gangetic plains but off late the crop has spread to Maharashtra, Karnataka, Andhra Pradesh, Madhya Pradesh, Uttar Pradesh and Gujarat. The area and production of pigeonpea in India have been growing at an annual rate of more than 2% since 1970. However, stagnant yields at around 700 kg ha-1 have been the major concern, though development of hybrid pigeonpea has raised new hopes for busting yield plateau.

The production enhancement through genetic amelioration has been a continuous process. Both conventional and modern plant breeding methods have been employed to breed improved types. Unlike other crops, pigeonpea gets bog down with increased input application. Whenever water is applied to pigeonpea vegetative growth is promoted with less flowering leading to less yield. This evinces that the plant type of pigeonpea is structured for evolutionary advantages like excessive growth, spreading habit, height, shattering, low harvest index and indeterminate growth with inapt ratio or balance of vegetative and reproductive phase. Thus, restructuring of plant type is the top priority. Breeding for appropriate spread, growth habit, maturity, height and characters as per requisite crop husbandry may serve the purpose.

Short duration varieties are known for their limited biological yield and further enhancement of harvest index would not guarantee high productivity in these cultivars. Therefore, sufficiently large biomass is equally necessary under arid conditions. Efforts should be directed to increase biological yield with optimum harvest index in short duration cultivars. Wild species as potent tools for pigeonpea improvement for arid situations: Wild relatives of pigeonpea are expected to have accumulated useful genes to tolerate droughts, floods, extreme temperatures, insect pests and diseases that wreak heavy damage to cultivated species. The natural defense mechanisms in crop plants have been lost during intense selection for high yield, wider adaptability and improved nutritional quality. The arid situations being very harsh, the use of mining these important genes for enhancing yield potentials can not be over emphasized. Discovery and incorporation of genes from wild species provide means to sustain crop improvement. Exploitation of cytoplasm from compatible wild germplasm has resulted in development of new cytoplasmic male sterility systems in pigeonpea, development of high protein lines, cleistogamous flower and dwarf pigeonpea lines. Utilization of wild species in secondary and tertiary gene pools has been generally restricted due to sterility, constrained recombination or cross incompatibility.

Nevertheless, these species are extremely important as they contain high levels of resistance to several important biotic and abiotic stresses. Breaking yield plateau through heterosis breeding: Pure line pigeonpea varieties released for commercial cultivation have crossed a century mark ever since the first pigeonpea variety was developed by selecting a wilt resistant genotype from landraces. These concerted research and developmental efforts have made substantial dent in pigeonpea production and areas, but the productivity has remained by and large static, despite various breeding methods such as pure line breeding, population breeding and mutation breeding.

In fact, a hybrid breeding system using partial natural out-crossing of the crop coupled with genetic male-sterility was tried to break yield stagnations by Reddy *et al.*, 1978 and Saxena *et al.*, 1983. Combining these two components, Saxena *et al.*, 1992 developed the world's first pigeonpea GMS based hybrid ICPH 8 that was released for cultivation in 1991 which recorded an average increase of 30.5% in grain yield over the best available pure line variety UPAS 120 (Saxena *et al.*, 1992).

Despite these achievements in genetic male sterility based hybrids in 90s to bust the yield plateau, they could not be exploited commercially due to inherent seed production problem with the system. This necessitated breeding for a more efficient cytoplasmic nuclear male-sterility (CMS) system. The pigeonpea breeders took up this mission and the first report of developing stable, versatile and commercially viable alternative cytoplasmic genetic based male sterility system using wild species *C. scarabaeoides* came from Sardarkrushinagar (Gujarat). The plant to plant selfing and back crossing of the sterile and fertile counterparts culminated in the development of stable cytoplasmic male sterility system in 1997. The system became viable with the identification of fertility restorer genes and first ever CGMS based hybrid viz. GTH 1 was identified in 2004. Appreciating that genetic diversity is the backbone of any successful hybrid development programme; the nation wide successful efforts were aimed at diversification of A and R lines. The success of the programme can be gauged from the fact that as on day Sardarkrushinagar (Gujarat) alone has identified and characterized 98 stable CGMS lines and 81 stable restorer lines besides host of other materials in pipeline.

The research conducted so far in this aspect has used the following five CMS cytoplasms (A1 through A5) conditioning male sterility. Broadening of narrow genetic base: Broadening of narrow genetic base makes an important activity for breeding under arid situation so that the varieties so developed may withstand the vagaries of the harsh weather. This has been successfully attempted through use of both indigenous and exotic germplasm.

Lately exotic lines from Kenya (ICP-9140 and ICP-9135), Tanzania (ICP-12116 and ICP-12161), Myanmar (ICP-11488) and Canada (ICP-13555) and diverse indigenous genotypes (GT-100, GT-101, Banas and ICP-11912) were used to develop different types of derivatives. The exotic germplasm though has very long duration but evinced immense complimenting ability of variability to Indian counterparts and, therefore, seemed very useful both for broadening the narrow genetic base and providing good combination for adequate exploitation of yield components and heterosis in pigeonpea.

The studies indicated that ombinations of Indian pigeonpea as female with exotic germplasm, particularly the African as male gave better specific combinations for yield per se.

***Abiotic Stress Resistance:*** Drought and soil salinity are the indispensable abiotic stresses in arid areas where pigeonpea is produced. Consequent up on lack of appropriate screening technologies and drought being highly complex and stage specific in nature, no systematic breeding program has been undertaken to breed varieties with high levels of resistance to these stresses. Though pigeonpea has deep roots, yet yield losses due to drought are large and widespread, especially when it occurs during critical seedling and reproductive stages. Lopez *et al.*, 1996 reported genotypic differences for drought resistance in pigeonpea. Such variation was associated with the ability of resistant genotypes to maintain efficient production and partitioning of dry matter during periods of drought. Chauhan *et al.*, 1994 identified a promising short duration cultivar ICPL 88039 that showed a moderate level of drought resistance.

Consequent upon strong relationship between the yield potential and the sensitivity of genotypes to end-of-season drought, the first approach to minimize yield losses due to terminal drought has been to breed for earliness for escaping drought. However, drought resistance being growth-stage specific such relationships do not hold true for mid season drought there by necessitating specific genetic enhancement programmes for drought resistance.

Little effort has been made to develop lines with genetic resistance against salinity stress in pigeonpea. A pigeonpea mini-core collection along with some wild species and field collections from predominantly saline areas was evaluated for resistance to irrigation with 75 mM NaCl solution (Srivastava *et al.,* 2006).

***Dry matter Production and Harvest Index:*** Pigeonpea produces the high dry matter though only a part of it is converted to economic yield. Exotic cultivars produced more biomass *i.e.,* more branches and leaves and had bushier and more compact canopy structure than the local genotypes. Pod yield was higher in the local genotypes than that in exotic cultivars (Fakir *et al.,* 1992). Dry matter production is also affected by environmental factors. Harvest index in pigeonpea varies between 10 and 52% depending upon genotypes, environments and agronomic managements. Plant breeder should aim such a plant type where the canopy closure is achieved prior to pod filling to ensure maximum interception of incident, photo-synthetically active radiation during reproductive ontogeny, but after start of flowering, to avoid excessive vegetative growth. The manipulation of photo-thermal sensitivity offers the most powerful tool for improving harvest index.

***Genetic Transformation:*** Judicious application of biotechnological tools holds great potential for alleviating some of the major constraints to productivity. The genetic engineering of pigeonpea could provide an effective complementary approach to *Helicoverpa* control. Several insect resistance genes have been identified as candidates for transforming pigeonpea including the genes coding for the insecticidal crystal protein of Bt, soybean trypsin inhibitor (SBTI) and pigeonpea trypsin inhibitor (PPTI). *35S:cry1Ab, 35S:sbti,* 35S:*cry1Ac, Ubi:cry1Ac* have been introduced successfully in pigeonpea against *Helicoverpa armigera,* which were being tested in contained field at ICRISAT and at present various events are available. For Bio-fortification for the nutritional components like ß-carotene, Sulphur-rich amino acids (methionine), the genes namely *oleo:psy* and *Vicilin: SSA* have been transferred to pigeonpea and the events being tested are available. Genetic engineering approaches to introduce agronomically desirable genes from other sources into pigeonpea has, so far not been possible due to the want of an efficient and reproducible regeneration system. An effort to develop efficient regeneration procedures a prerequisite to plant improvement several regeneration protocol via direct and indirect organogenesis or embryogenesis using diverse explants on different media have been established. However, the regeneration frequency is very low dependent on the explant nutritional and hormonal regimes and genotype specificity. Regeneration of plants from protoplast, anther and rescued embryo cultures have been achieved with limited success. Very few studies have so far been undertaken for genetic transformation in pigeonpea.

Development of primary transformants expressing reporter and selectable marker genes have been achieved at very low frequency using direct shoot organogenesis from cotyledonary nodes and *Agrobacterium tumefecians*. However, a protocol which can routinely be used for genetic transformation in pigeonpea is still far from being reached (Singh *et al.*, 2003) Large scale development of microsatellite markers has been initiated in pigeonpea. Two enriched libraries were created resulting in low recovery rate of only 10 useful microsatellite markers. Due to the substantial amount of time and cost incurred, there is considerable interest in establishing a more cost effective and labour saving alternative in generating additional markers for drought tolerance.

Marker-assisted selection: In most breeding programmes, the genetic improvement for drought resistance is accomplished through selection for yield and because of low heritability of yield under stress and the spatial as well as temporal variation in the field environment, conventional breeding approaches are slow. Whereas molecular markers such as restriction fragment length polymorphism (RFLP), random amplified polymorphic DNA (RAPD) and isozyme will facilitate to develop drought-resistant genotypes more effectively as their expressions are independent of environmental effects.

Until recently, there have been no microsatellite markers available in pigeonpea. However, a collaborative project between ICRISAT and Birmingham University is now developing SSR markers that have already been used in the first high throughput project at ICRISAT, to assess the molecular diversity of one thousand pigeonpea accessions from across India. Meanwhile, a number of components of drought tolerance have been characterized in pigeonpea, including early vigor, leaf area maintenance, root and shoot growth rate, and developmental plasticity.

Molecular markers are also important for investigating gene flow between crops and their wild species. Work has been initiated by many institutions for the development of a model for gene flow in pigeonpea, using DNA markers, to understand the risk of genetic erosion as well as to have a tool for risk assessment when transgenic plants of this crop become available for field-testing.

Seed production and maintenance breeding: The maintenance of genetic purity is quite arduous in crops like pigeonpea owing to its often cross pollination nature. The situation cannot be more warranting than crop improvement programme in general and hybrid programme in particular where a single gene migration can set in enormous genetic upheaval for attaining gene and gene frequency equilibrium. A simple but very effective technology for maintenance of genetic purity was designed at Sardarkrushinagar (Gujarat). The technology comprises 8 feet iron poles with detachable top. The net is erected on the arches made of poles and tops. The technology was further improved and presently the mobile insect proof net houses are being used country wide for maintaining genetic purity and maintenance breeding.

## Researchable Issues in Arid Regions

- Pigeonpea in arid region has special niches in marginal and submarginal land under rainfed areas. Therefore, development of short duration varieties/hybrids with appropriate vegetative growth, suitable spread, balance between vegetative and reproductive phase, least bud dropping, consumers' friendly quality, resistance to biotic and abiotic stress with due emphasis on anti-evolutionary characters like nonshattering, may be targeted to increase and stabilize pigeonpea production under arid regions.
- Development of situation specific and physiologically efficient varieties.·
- Utilization of wild relatives for introgression of drought resistance, high protein and pest/disease resistance genes in to cultivated pigeonpea.
- Development of value added dual purpose varieties with good quality attributes.
- Tolerance to biotic and abiotic stresses for escalating and stabilizing production over locations and years under arid conditions with particular emphasis on the development of drought tolerant varieties with due emphasis on photo-thermo-insensitivity.
- Development of molecular based mechanism and techniques to enhance for drought, pest/disease resistance.
- Incorporating resistance to Helicoverpa.
- MAS could be resorted for selection for resistance to drought and salinity.

- Genes for higher harvest index may be exploited in pigeonpea for increasing the harvest index and yield.
- Development of character based specific molecular techniques for DUS and precise identification of varieties/hybrids for GOT at juvenile stage.

## Sesame

Sesame is the oldest indigenous oil crop cultivated in India for its excellent nutritional, medicinal, cosmetic and cooking qualities of oil. It is a rich source of protein (24%) and carbohydrates (15%) in addition to excellent source of quality oil (50%). India is the world leader with the largest area (29.3%), maximum production (25.8%) and highest export of sesame seed (40%). Rajasthan, Gujarat, Utter Pradesh, Madhya Pradesh, Maharashtra, Andhra Pradesh, Orissa, Tamil Nadu and Karnataka are sesame growing states of India.

Among these, Rajasthan is the major sesame growing state in the country, which contributes 18.2% in area and 14.1% in production. The productivity trend (254 kg ha-1 to 356 kg ha-1) of the state during the past decade (1990-2000) as compared to productivity of 146 kg to 300 kg ha-1 during 1970-90 reflects the contribution of research in development of drought, disease and pest tolerant and high yielding varieties adopted by farmers. But still productivity level (327 kg ha-1) of the state is low in comparison to the national productivity of 421 kg ha-1. There are several constraints in sesame production as it is mainly grown in the *kharif* season under vagaries of monsoon, which is further aggravated by several pests and diseases. Recently prospects of export of white seeded sesame have brightened and Rajasthan is contributing greatly in its export. However to meet the growing demand, the yield ceilings of sesame need to be raised by developing high yielding varieties having tolerance to biotic and abiotic stresses. Not only the higher productivity but desired traits like whiteness of seed, technology for raising pesticide residuefree and organically produced crop also needs due attention.

Systematic and scientific efforts to increase the production and productivity of this crop in the Rajasthan state started in early 1970s and since then several varieties have been released. Rajasthan Til 46: It is a white seeded high yielding variety resistant to oozing complex, tolerant to *Macrophomina, Alternaria* leaf spot, *Antigastra* and gall fly. It matures in 76 to 85 days and is suitable for both rainfed as well as irrigated conditions. Its seed yield ranges from 600-800 kg ha-1 and has 49.4% oil content. It is recommended for cultivation in J & K, Himachal Pradesh, Punjab, Haryana, Rajasthan and Western U.P.

Rajasthan Til 54: A brown seeded and comparatively dwarf variety which is resistant to leaf blight and Macrophomina. It is comparatively late shattering type and matures in 73-78 days. Seed yield ranges from 6-8 q ha-1 under natural conditions and is recommended for cultivation in Gujarat, Maharashtra and Telangana region of A.P. Rajasthan Til 103: A white seeded variety suited for both rainfed and irrigated conditions, highly tolerant to bacterial leaf blight and insects, is a late shattering type variety with seed yield of 600-800 kg ha-1 under rainfed conditions. It is recommended for Gujarat, Maharashtra and Telangana region of A.P.

Rajasthan Til 125: It is characterized by yellowing of leaves, stem and capsules as the crop reaches physiological maturity. It is tolerant to diseases and insects and gives seed yield of 600-800 kg ha-1 under rainfed conditions and has 48.7% oil content. It is recommended for cultivation in J & K, H.P, Punjab, Haryana, Rajasthan and Western U.P. Rajasthan Til 127: A shiny white bold seeded variety for cultivation in whole Rajasthan state. It is drought hardy and well suited for both arid & semi arid conditions. It is tolerant to Macrophomina stem & root rot, phyllody, bacterial leaf spot & powdery mildew and has less infestation of gall fly and mites. It is also tolerant to moisture stress conditions.

Its average seed yield is 6-9 q ha-1 and oil content 50.6%. Rajasthan Til 346 (Chetak): It is a white seeded, early maturing (83 days) and high yielding variety. It has 3-4 branches, non hairy capsules arranged alternatively in a compact manner. It is resistant to leaf curl, phyllody and moderately resistant to *Macrophomina* stem and root rot, powdery mildew and *Alternaria* and *Cercospora* leaf spot and has less infestation against *Antigastra*. This variety also showed tolerance against moisture stress. Its average seed yield is 700-900 kg ha-1 and oil content 50%. It is suitable for cultivation in hot and arid ecosystem of the National zone-I, comprising of Rajasthan, Haryana, Punjab, Himachal Pradesh, Gujarat and adjoining areas of western Uttar Pradesh, Maharashtra and Karnataka. Impact of sesame varieties developed: Variety RT 46, RT 125 and RT 127 have already become popular in the Rajasthan state. In front line demonstrations conducted in western Rajasthan varieties RT 46, RT 125 and RT 127 have exhibited 35-40% higher seed yield over local varieties. Seed demand of these varieties is continuously rising in the state due to its high realized productivity on farmers' field. Varieties RT 46, RT 125 and RT 127 are presently grown in more than one lakh hectare area, which is about 35% of the area under sesame in Rajasthan. With more support of state seed producing agency, the 50% area under these varieties can be attained in next 3-4 years.

## Rapeseed and Mustard

*Brassica* species is an important component of human diet as major sources of edible oil and vegetables. Seven important annual oilseeds are grown world wide. India has traditionally been a *Brassica* growing country. Three ecotypes of *B. rapa* namely brown *sarson*, yellow *sarson* and *toria* are grown to a limited extent while Indian mustard (*B. juncea*) is the chief oilyielding crop. Black mustard or *Banarsi rai* (*B. nigra* [L.] Koch) is cultivated for condiments in a few pockets. Two more oilseed *Brassica* species i.e. rapeseed (*B. napus*) and Ethiopian mustard (*B. carinata*) have been created. *B. tournefortii* is also grown sporadically in the drier habitats of Rajasthan on marginal areas.

The production of oilseed *Brassica* increased to 1 m t the 1956-57 to 7.5 m t in 2006-07 due to concerted efforts of breeding and crop management. Rajasthan has the largest acreage under these crops and is ranked first in area (3.21 m ha) as well as in production (3.81 m t) in the country and accounts for about 50% production. Low irrigation requirements of these crops compared to other *rabi* crops like wheat and barley has played an important role in popularizing the crops in Rajasthan where water is scanty.

Genetic diversity: The divergence studies are necessary to understand the trend of evolutionary pattern, to assess the relative contribution of different components of yield to total divergence and to determine the nature of forces operating at intra- and inter-cluster levels. Choudhary and Joshi (2001) assessed genetic diversity among 88 entries including eighty F4 derivatives of interspecific crosses and eight parents.

The genetic distances calculated among different *Brassica* species revealed that *B. tournefortii* had maximum diversity with *B. juncea* followed by *B. napus*, *B. rapa* var. *toria*, and *B. rapa* var. yellow *sarson*. The clustering pattern showed that many derivatives of the cross fell into the same cluster but in many cases in spite of common ancestry descendants of the cross spread over different clusters. The characters like plant height, secondary branches per plant, days to flowering and 1000-seed weight contributed maximum towards genetic divergence.

In another study of genetic divergence of 84 entries of *B. juncea* showed that all the 28 entries having tetralocular siliqua were accommodated in a single cluster, whereas remaining 56 genotypes with bilocular siliqua were spread over in five clusters (Choudhary *et al.*, 2007). The mean values of the traits in various clusters exhibited that different clusters were superior in respect of different characters. Contribution of different characters towards total genetic divergence revealed highest contribution of seeds per siliqua followed by main raceme length, siliquae on main raceme and siliqua length. Genetics of economic important traits: Numerous studies on genetics of quantitative and qualitative characters revealed availability of genetic variability in different *Brassica* species.

The inheritance of siliqua locule number and seed coat colour in *B. juncea* was investigated using three lines each of tetralocular brown seeded and bilocular yellow seeded (Choudhary and Solanki, 2007). Brown seed colour and bilocular siliqua characters were found to be dominant over yellow seed and tetralocular siliqua, respectively. Another study on the inheritance of siliqua orientation and seed coat colour in *B. tournefortii* showed that seed colour was to be under monogenic control with brown being dominant over yellow (Choudhary, 2008).

Crop improvement: Crop improvement objectives focus largely on attempts to tailor plants for higher seed yield, resistance/tolerance to pests and diseases, frost, shattering, and for oil and meal quality. The breeding approaches have led to the development of over 150 improved cultivars in last few decades resulting in higher productivity of crop. Pusa Jaikisan (Bio 902) of Indian mustard is the first somaclonal variety developed through tissue culture. With regards to quality, Hyola 401 is the first '00' hybrid of *B. napus* while Neelam (HPN 3) is the first '00' variety of the same species. PGSH 51 is the first CMS based hybrid of *B. napus*. Variety CS 52 has been recommended for salt affected soils. Indian mustard variety JM 1 is reported to be tolerant to white rust while variety RGN 13 is tolerant to frost and *Alternaria* blight. Some important varieties recommended for arid western plain zone Ia are Varuna, RH 30, RH 819, Pusa Jaikisan (Bio 902), CS 52, GM 2, JM 1, Urvashi and Ashirwad of Indian mustard and ITSA, T 27 and RTM 314 of taramira.

## Seed Spices

The seed spices are a group of annual crops, the dried fruit or seeds of which are used as spices. The seed spices are aromatic vegetable products of tropical origin and are commonly used in pulverized state, primarily for seasoning or garnishing the foods and beverages. They are also used in preparation of various value-added products viz. spice oils, oleoresins and spice powders. The area under seed spices in India is 9,22,274 ha and the production is 5,94,120 metric tonnes. India is the largest producer, consumer and exporter of seed spices earning Rs. 5440 millions by exporting 52,550 tonnes seed spices to various countries (Anonymous, 2009).

***Crop Improvement Achievements:*** After inception of AICRP on Spices in 1975, major achievements have been made and over 50 improved varieties in seed spices have been released. These include 19 in coriander, 7 in cumin, 10 in fennel, 16 in fenugreek, 7 in ajowan, 2 in dill and one each in nigella, celery, and anise. Different biotic and abiotic stresses are major yield reducing factors in seed spices. Therefore the breeding objectives for seed spices include resistance against these factors and also enhancement of nutritional, yield and quality characters like high volatile oil content.

***Coriander (Dhania):*** *Coriandrum sativum* L belongs to family apiaceae and requires improvement for high yield, higher volatile oil content, resistance to powdery mildew and stem gall disease and small and round seed with luster. So far several varieties viz., RCr-41, RCr- 20, RCr-435, RCr436, RCr-684, RCr-446, Ajmer Dhania-1 (NRCSS ACr 1) for Rajasthan; GCr-1 and GCr-2 for Gujarat; CO 1, CO 2, CO 3 and CS 287 for Tamil Nadu; Rajendra Swathi and Rajendra Sonia for Bihar; Sadhna, Swathi (CS-6) and Sindhu, for Andhra Pradesh and Hisar Sugandh for Haryana have been released.

***Cumin (Zeera):*** *Cuminum cyminum* L belongs to family apiaceae and requires improvement for resistance to blight, wilt and powdery mildew diseases, high yield, higher volatile oil content and bold seed with luster. So far varieties developed and released include RZ-19, RZ- 209 and RZ-223 for Rajasthan and GC-1, GC-2, GC-3 and GC-4 for Gujarat.

***Fennel (Saunf):*** *Foeniculum vulgare* Mill belongs to family apiaceae and requires improvement for short duration, reduced height, large umbel, high yield, resistance to gummosis and high volatile oil content. Varieties released for Rajasthan include Ajmer Saunf-1 (NRCSS-AF-1), RF-101, RF-125, RF- 143; for Gujarat include PF-35, GF-1 (Gujarat Fennel-1) GF-2 and GF-11. Variety Co-1 is released for Tamil Nadu and Hisar Swarup for the state of Haryana.

***Fenugreek (Methi):*** *Trigonella foenum-gracium* L belongs to family fabaceae and requires improvement for high yield, resistance to powdery and downy mildew, earliness and improved medicinal properties. The varieties RMt-1, RMt-143, RMt-303, RMt-305, Ajmer Methi 1 (NRCSS-AM-1) and Ajmer Methi 2 (NRCSS-AM-2) are recommended for Rajasthan; GM-1, for Gujarat; CO-1 for Tamil Nadu; Rajandra Kranti for Bihar; Lam Selection–1 for Andhra Pradesh; Hisar Sonali, Hisar Suvarna, Hisar Mukta, Hisar Madhavi (HM-350) for Haryana; Pant Ragini, Pusa Early Bunching, Pusa Kasuri for Uttar Pradesh and Uttaranchal.

***Dill (Sowa):*** *Anethum graveolens* L European dill and *Anethum sowa* Kurk. belongs to family apiaceae and requires improvement for high yield, higher volatile oil content, resistance to powdery mildew and root rot disease. Ajmer Sowa 1 (NRCSS-AD-1) and Ajmer Sowa 2 (NRCSS-AD-2) varieties have been recommended for Rajasthan.

***Carom Seed (Ajwain):*** *Trachyspermum ammi* L (Sprague) belongs to family apiaceae and requires improvement for high yield, higher volatile oil content, resistance to powdery mildew and root rot disease. The suitable varieties are Ajmer Ajwain 1 (NRCSS-AA-1), Ajmer Ajwain 2 (NRCSS-AA-2), Pratap Ajwain 1 for Rajasthan; GA-1 for Gujarat; Lam Selection-1, Lam Selection–2 for Andhra Pradesh, R.A. 1-80, R.A. 19-80 for Bihar.

***Nigella (Kalongi):*** *Nigella sativa L.,* belongs to family *Ranunculaceae* and requires improvement for high yield, higher volatile oil content. Ajmer Kalongi 1 (NRCSS AN-1) is suitable for cultivation in semi-arid regions under irrigated conditions.

***Anise (Vilayati Saunf):*** *Pimpinella anisum L.,* belongs to family apiaceae and requires improvement for high yield, higher volatile oil content and resistance to powdery mildew. The variety Ajmer Vilayati Saunf 1 (NRCSS AAni-1) is a high yielding variety suitable for cultivation in semi - arid region under irrigated conditions.

***Celery (Ajmud):*** *Apium graveolens L.,* belongs to family apiaceae and requires improvement for high yield. The variety Ajmer Ajmud 1 (NRCSS A Cel-1) is suitable for cultivation in semi-arid regions under irrigated conditions.

## Future Strategy for Crop Improvement

1. Serious attention must be paid to the systematic germplasm collection, evaluation and conservation. The prevailing diversity of these crops in remote and tribal pockets and landraces be tapped.
2. Introduction of exotic material from centre of origin and secondary centre of variability.
3. Molecularly characterization of all the germplasm with associated specific characters.
4. Free exchange of germplasm among the breeders involved in the improvement programme needs to be encouraged.
5. Mutation and biotechnological approaches to enhance the variability may be accelerated.
6. Population improvement vis-à-vis selection to be carried out.
7. Detailed studies need to be carried out on floral biology, anthesis, pollination and male sterility – both natural and induced.
8. Heterosis breeding to be initiated.

9. Breeding for disease and pest resistance, quality, stability and responsiveness to agronomical inputs.
10. Breeding varieties tolerant/resistant to abiotic stress of temperature, moisture, salinity, etc.

## Range Grasses and Legumes

In the majority of pasture grasses genetic improvement through hybridization is rather difficult due to pre-dominantly apomictic mode of reproduction. Some of these grasses are facultative apomictic with varying degrees of sexuality whereas, a few others are obligate apomictic. Apomixis coupled with polyploidy causes a lot of problems in grass variety improvement programme. Further, the minute floral parts, self-sterility, plurality of species, poor seed production and complex evaluation procedures make the task of the breeder more difficult. In grassland farming, the compatibility of genotypes with other plant types is more relevant. Not all the species are compatible to each other and this sometimes creates problems. Suitable compatible species need to be grown in the pasture for increased production.

*Lasiurus sindicus* and *Cenchrus ciliaris* are more compatible with legume stylo than other legumes. In winter cropping system, suitable ideotype and their use becomes more relevant in formulating compatible mixture of grasses and legumes. The specific growth forms of grasses either artificially tailored or selected from natural source account for better production per unit area and time even with similar input levels. Although their ecological niche governs the occurrence of species, which takes into account various biotic and abiotic factors. The improvement of pasture species in a particular agro-climatic zone is of vital importance. Keeping in view the annual precipitation, its distribution, soil types, topography and temperatures, etc. the choice of suitable species is the kingpin of a pasture improvement programme. In pasture grasses the breeding system must be determined before undertaking any breeding programme. Simple tests using porous cellophane or nylon fabric bag will indicate whether a species is self or cross-pollinated.

Good seed set from bagging a single inflorescence generally indicate selfpollinated species, but in grasses it could be due to apomixis. Progeny testing and embryological studies may be necessary to distinguish between apomictic and sexual grasses. Many important tropical grasses are apomictic including buffel grass, birdwood grass and *Panicum maximum*. Apomixis is a barrier to improvement of these grasses and can be overcome only if sexual types are found. Sexual types are found in *C. ciliaris* (Bashaw, 1962) and *P. maximum*.

Most of the sexual tropical grasses viz., *Setaria* and *Digitaria* are predominantly cross-pollinated and self-sterile. In making crosses between ecotypes it is wise to check first that the representative genotypes are selfincompatible.

The critical information in grass improvement includes magnitude of genetic variation, stability and polymorphism. Genetic variability: Variability in morphological and physiological characters in 10 strains of *Cenchrus setigerus* showed that some of the strains could provide useful germplasm to breed desirable forage type for pasture production (Chakravarty and Bhati, 1968). Wide genetic variation was observed in the characters contributing to forage yield, forage quality and underground biomass for *C. ciliaris, C. setigerus, L. sindicus, Dichanthium annulatum* and *Pennisetum pedicellatum* (Singh and Yadav, 1971; Yadav *et al.*, 1980; Yadav and Krishna, 1985; Yadav and Krishna, 1987; Rajora, 1998).

The extent of variability, heritability estimates and genetic advance were high for plant height, number of spikes/plant, spike length, seeds/spike, 1000-seed weight, seed germination and seed yield in winter regenerated flush of *C. ciliaris* genotypes (Rajora and Singh, 2005). A wide range of variation was observed for growth habit, plant pubescence, spike colour, stigma colour root rhizome production in *C. setigerus* (Yadav, 1981). In a three year study Rajora *et al.* (2009) observed that in *C. setigerus* var.

CAZRI 76 maximum variation in seed yield was due to spike density and seed yield could be harvested from the established plots up to 3 years under rainfed arid environmental conditions. High coefficient of variation, moderate to high heritability and high genetic advance were observed in *C. ciliaris, L. sindicus* and *D. annulatum* for tiller number, stem thickness and fodder yield. A wide range of variation was observed in *L. sindicus* for quality traits, viz. dry matter, silica, lignin and crude protein content. These characters showed high heritability and high genetic advance (Gupta and Yadav, 1985). Significant genotypic differences were observed in ten genotypes of *L. purpureus* for seed yield per plant, seeds per pod, 100-seed weight, pod length, pod weight, seed weight per pod, dry matter yield per plant and seed germination (Rajora and Yadav, 1999). Stability: Sharma (1985) evaluated eight cultivars of *Stylosanthes* species received from Australia. Preliminary studies under irrigated conditions revealed that their germination was low; some of the cultivars did not establish and set seed. *Stylosanthes hamata* C.P.I. 38842 and *S. scabra* were superior in forage yield and seed production. Maximum dry forage yield (2.71 t ha-1) was obtained from *S. scabra*. C.P.I. 40205. Suitable harvest time for maximum forage yield was found to be December.

Promising strains of *C. ciliaris, C. setigerus, Pennisetum pedicellatum, L. sindicus, Dichanthium annulatum* and *Lablab purpureus* were studied under different locations, years, soil types, fertility, defoliation stress, photoperiods and grazing (Yadav, 1981; Roy *et al.*, 1995). Four genotypes of *C. ciliaris* differed significantly for plant height, spikes/plant, dry matter yield/plant, dry matter yield, seed yield/plant, seed yield and seed germination. CAZRI 358 and extant variety having maximum spikes/plant (100.4), seed yield/plant (20.74 g), seed yield ha-1 (266.0 kg) with high germination (44.1%), was more suitable for seed production in arid areas (Rajora *et al.* 2008). Development of improved varieties and genetic stocks: Variety CAZRI 75 of *Cenchrus ciliaris* developed at CAZRI is tall (90-125 cm), erect, thick stemmed, leafy, drought hardy, with green yield potential of 7 t ha-1 year-1 yielding 53% higher over the check var. Molopo. It has high crude protein (8.3%), which is 32% higher over the check (Yadav and Sharma, 1998).

The variety CAZRI 76 of *Cenchrus setigerus* is a clonal selection from exotic material EC 17655, medium tall (50-60 cm), thin stemmed, leafy (leaf-stem ratio 2.3) foliage remains green up to December. It is drought hardy, high tillering ability (29 tillers plant-1), m fast regeneration (2-3 cut in a year), average green fodder yield (4 t ha-1) and 1.5 t dry ha-1 matter which is 38% higher over the check var. Pusa Yellow Anjan. It is also nutritious, having 9.6% crude protein at 50% flowering stage (Yadav and Sharma, 1998).

Accession CAZRI 2221 of *C. ciliaris* was identified as an *elite* genotype. It is semi-erect with has luxuriant growth, profuse tillering and branching, pink and smooth nodes, thin stem, yellowish foliage, longer leaves, high leaf: stem ratio (> 57%) than the best check. Long creamish spikes, plant height 100-105 cm. Green forage yield potential was 24 t ha-1, which was 20% more than the best check. It has high survival ability under arid conditions (>6-year) and IVDMD (61.9%). The accession CAZRI 2178 had overall 15.2% more green fodder yield than the best check CAZRI 75, 59% more CP than CAZRI 75, higher survival (89.3% in 3rd year), erect, and thin stem. CAZRI 2178 can be used as a genetic stock for the improvement programme.

# Bibliography


Alfred Steferud: *Diseases of Fruits and Nuts*, Biotech Books, Delhi, 2005.

Anderson, E. F.: *The Cactus Family*, Timber Press, Portland, Oregon, 2001.

Andrew L. Winton and Kate Barber Winton : *A Handbook of Structure and Composition of Foods*, Agrobios Publication, Delhi, 2013.

Andrews L.: *Citrus Production - Orange*, St. Augustine, Trinidad and Tobago, 1990

Arora, Chandralekha : *Advances in Horticulture Science Research*, Oxford Book Company, Delhi, 2009.

Ashok K Chauhan and Ajit Varma:*Microbes : Health and Environment*, I K International, Delhi, 2007.

Aysha Aziz Faridi:*Dairy Microbiology*, Random Publications, Delhi, 2012.

Benson, L. H.: *The Cacti of the United States and Canada*. Stanford University Press, Stanford, California, 2001.

Benu Singh: *A Modern Book on Forestry and Horticulture*, Vista International Pub, Delhi, 2010.

Bhandari, M.M.: *The Flora of the Indian Desert*. Scientific Publisher, Jodhpur, 1990.

Bhattacharjee, Supriya Kumar: *Advances in Ornamental Horticulture*, Pointer, Delhi, 2006.

Bhutani, R.C.: *Fruit and Vegetable Preservation*, Biotech Books, Delhi, 2003.

Chadha, K.L. : *Advances in Horticulture*, Malhotra Publishing House, Delhi, 2006.

Collymore L.: *Fruit Production in Barbados*, Port of Spain, Trinidad and Tobago, 1996.

Cristobal Noe Aguilar; Juliana Morales Castro; Efren Delgado; Diana Jasso Cantu and Ashok Pandey: *Food Science and Food*

Dabholkar, A.R. : *General Plant Breeding*, Concept, Delhi, 2006.

Dalip Kumar and Asmi Raza: *Agriculture and Food Security: Contemporary Issues*, Deep and Deep Publications, Delhi, 2011.

Doijode S. D.: *Seed Germination in Fruits*, New Delhi, Malhotra Publishers, 1993.

Doreen Virtue : *Constant Craving : What Your Food Cravings Mean And How To Overcome Them* , Hay House India, Delhi, 2011.

Goswami, C L : *Plant Physiology : Research Methods*, Scientific, Delhi, 2007.

Harmeet Singh: *Dairy Farming*, APH Publication, Delhi, 2011.

Hema Thapar: *Food Science and Health*, Pacific Publications, Delhi, 2011.

Julia F.: *Fruits of Warm Climates*, Miami, Julia F. Morton Publisher, 1987.

Kumar Supriya : *Amenity Horticulture, Biotechnology and Post-Harvest Technology*, Pointer, Delhi, 2006.

Little, E.L.: *Common Fuelwood Crops*, Communi-Tech Assoc., Morgantown, West Virginia, 1983.

Madhulika Bhatnagar.: *Encyclopaedia of Catering Technology, Food Service and Hospitality Management, Vol. I to III*, Delhi, Anmol Publication, 2008.

Malins A.: *Postharvest Handling of Pineapple and Mango*, Port of Spain, Trinidad and Tobago, 1992.

Manju Malhi: *Easy Indian Cookbook: The Step-By-Step Guide to Deliciously Easy Indian Food at Home*, Viva Books, Delhi, 2008.

Manoj Negi: *Career in Dairy Farming*, Abhishek Publication, Delhi, 2010.

Mitra S.: *Postharvest Physiology and Storage of Tropical and Subtropical Fruits*, Oxon, CABI, 1997.

Morton, Julia F.: *Fruits of Warm Climates*. Wipf and Stock Publishers, 1987.

Pareek O. P.: *Advances in Horticulture: Fruit Crops*, New Delhi, Malhotra Publishing House, 1993.

Paul B. : *Food Biotechnology in Ethical Perspective*, Aspen, CO: Aspen Publishers, 1997.

Peter, K.V. : *Basics of Horticulture*, New India, Delhi, 2009.

Piyush Sharma and Prem Ram : *A Textbook of Food and Beverage Management*, Dominant Publication, Delhi, 2013.

Popenoe, W.: *Manual of Tropical and Subtropical Fruits*. Hafner Press, 1974.

Proctor F. J.: *Onions in Tropical Regions*, Kent, Natural Resources Institute, 1990.

Purohit, S.S. : *Organic Farming : Biocontrol and Biopesticide Technology*, Agrobios, Delhi, 2008.

Saini, M.L. : *Plant Breeding and Crop Improvement*, CBS, Delhi, 1997.

Saroj, P.L. : *Advances in Arid Horticulture*, International Book Dist, Delhi, 2004.

Sharma, V.K. : *Advances in Horticulture : Strategies, Production, Plant Protection and Value Addition*, Deep and Deep, Delhi, 2004.

Singh Z.: *Propagation and Rootstocks of Citrus*, New Delhi, Malhotra Publishers, 1993.

Tapeshwar Singh: *Resource Conservation and Food Security : An Indian Experience (2 Vols-Set)*, Concept Publication, New Delhi, 2004.

Tharakan : *A Young Hotelier's Guide to Food and Beverage*, Tata McGraw-Hill Publication, Delhi, 2005.

Thomas E.: *Fruit Production in St. Kitts and Nevis*, Port of Spain, IICA, 1996.

Tyagi, I.D. : *Plant Breeding and Genetics at a Glance*, South Asian, Delhi, 2005.

Udai Veer: *Elements of Food Science*, Anmol Publications, Delhi, 2007.

Urvashi Nandal : *A Handbook of Foods and Nutritional Biochemistry*, Agrobios Publication, Delhi, 2013.

Vishwambhar Prasad Sati: *Natural Resources Conservation and Food Security*, Bishen Singh Mahendra Pal Singh, Dehra Dun, 2012.

# Index

## R



## S



## T



## V



❑❑❑